The Open University

MU120
Open Mathematics

Unit 10

Prediction

MU120 course units were produced by the following team:

Gaynor Arrowsmith (Course Manager)
Mike Crampin (Author)
Margaret Crowe (Course Manager)
Fergus Daly (Academic Editor)
Judith Daniels (Reader)
Chris Dillon (Author)
Judy Ekins (Chair and Author)
John Fauvel (Academic Editor)
Barrie Galpin (Author and Academic Editor)
Alan Graham (Author and Academic Editor)
Linda Hodgkinson (Author)
Gillian Iossif (Author)
Joyce Johnson (Reader)
Eric Love (Academic Editor)
Kevin McConway (Author)
David Pimm (Author and Academic Editor)
Karen Rex (Author)

Other contributions to the text were made by a number of Open University staff and students and others acting as consultants, developmental testers, critical readers and writers of draft material. The course team are extremely grateful for their time and effort.

The course units were put into production by the following:

Course Materials Production Unit (Faculty of Mathematics and Computing)

Martin Brazier (Graphic Designer)
Hannah Brunt (Graphic Designer)
Alison Cadle (TEXOpS Manager)
Jenny Chalmers (Publishing Editor)
Sue Dobson (Graphic Artist)
Roger Lowry (Publishing Editor)
Diane Mole (Graphic Designer)
Kate Richenburg (Publishing Editor)
John A.Taylor (Graphic Artist)
Howie Twiner (Graphic Artist)
Nazlin Vohra (Graphic Designer)
Steve Rycroft (Publishing Editor)

This publication forms part of an Open University course. Details of this and other Open University courses can be obtained from the Student Registration and Enquiry Service, The Open University, PO Box 197, Milton Keynes MK7 6BJ, United Kingdom: tel. +44 (0)845 300 6090, email general-enquiries@open.ac.uk

Alternatively, you may visit the Open University website at http://www.open.ac.uk where you can learn more about the wide range of courses and packs offered at all levels by The Open University.

To purchase a selection of Open University course materials visit http://www.ouw.co.uk, or contact Open University Worldwide, Walton Hall, Milton Keynes MK7 6AA, United Kingdom, for a brochure: tel. +44 (0)1908 858793, fax +44 (0)1908 858787, email ouw-customer-services@open.ac.uk

The Open University, Walton Hall, Milton Keynes, MK7 6AA.

First published 1996. Second edition 2008.

Edited, designed and typeset by The Open University, using the Open University TEX System.

Printed and bound in the United Kingdom by The Charlesworth Group, Wakefield.

ISBN 978 0 7492 2868 2

2.1

Contents

Study guide

This is the first of a block of units which look at change. In particular, this unit is concerned with making predictions. The unit relies very much upon the work you did in Block B, especially the calculator, graphical and algebraic work. You will need to use graph paper as well as your calculator to draw graphs. This unit is a little longer than usual, but *Unit 11* compensates for this by being a bit shorter.

Section 1 discusses the idea of mathematical modelling as a representation for a purpose—an idea that you have met before. Section 2 concentrates on models based on a constant rate of change, which lead to straight-line graphs and their associated algebraic equations. Much of this is revision of your work on graphs and symbols from Block B (*Units 7* and *8*). Section 3 is based on the *Calculator Book*. When you have data to which you want to fit a linear (straight-line) model, then your calculator has a useful facility for doing this. Section 4 looks at situations involving *two* linear models—for example, predicting where two ferries or two trains will pass. You have met this type of problem before and have used your calculator to find where the graphs intersect; in this section you will use algebra to solve such problems.

Section 5 looks at the use of mathematical techniques in predicting optimal strategies. This section also involves work from the *Calculator Book*, and includes the optional study of an audio band. If you are short of time, you can skim through the section, although this would probably mean that you could not do a small part of the assessment.

This unit builds on material that you have learned in earlier units, and so provides a good opportunity for you to consider how you integrate familiar ideas with new ones. The skills and strategies that enable you to do this are sophisticated. One aspect of your Learning File work for this unit will be to think about this in your learning of mathematics.

The Handbook theme for this unit relates to mathematical modelling and the different terms you will be using. You should build up a comprehensive Handbook activity sheet on modelling and related terminology as you work through the text.

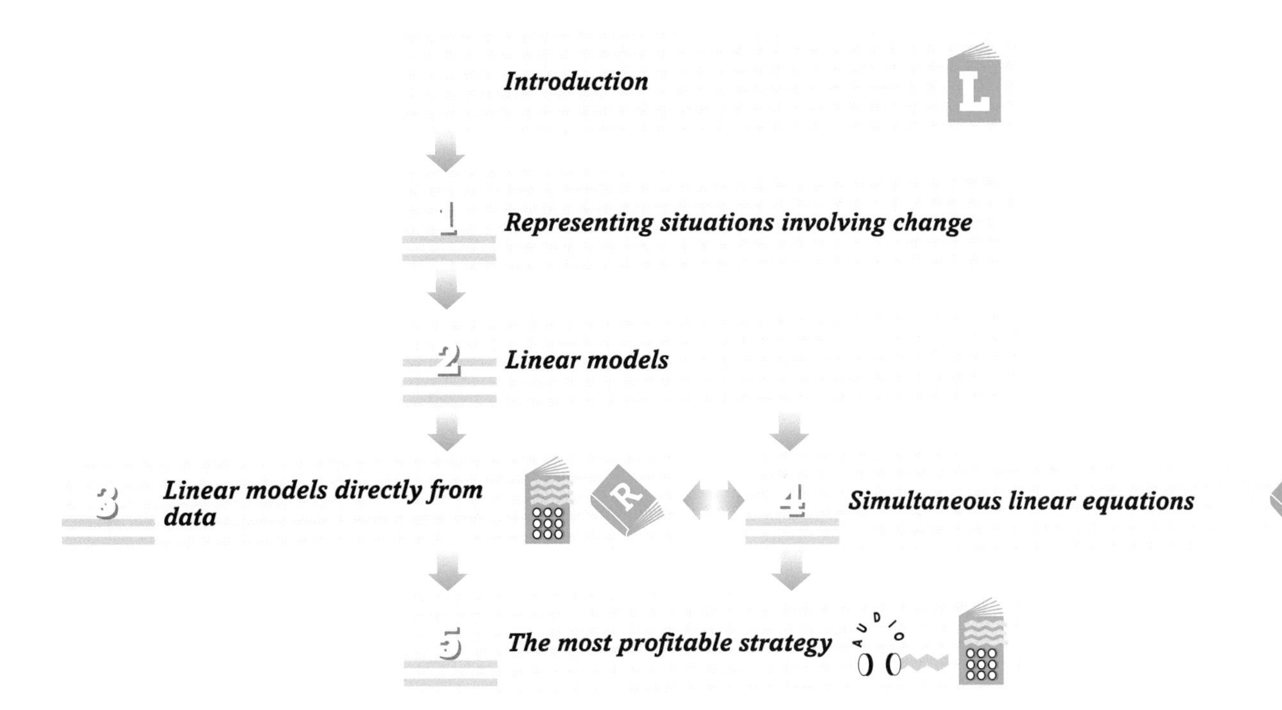

Summary of sections and other course components needed for *Unit 10*

Introduction to Block C

Many quantities change with time: the position of a moving vehicle, the radioactivity emitted from a piece of radioactive material, the value of capital in a savings account. However, the way many quantities change depends upon variables other than time: the quantity of a foodstuff bought may depend upon the price, the cooking time for a cake or joint of meat depends upon the weight. This block is about how mathematics can help in the understanding of how quantities change and in making predictions about such change.

Block A introduced the statistical investigation cycle: posing the question, collecting the data, analysing the data and interpreting the results. Block B introduced the idea of modelling as a representation for a particular purpose: stressing some aspects and ignoring others. Block C pulls these ideas together in modelling how things change. It also builds on the algebraic and graphical skills you learned in Block B and on the data-handling skills you learned in Block A in showing you how to use the course calculator to fit algebraic functions to given sets of data and how the idea of gradients of graphs is linked to the idea of rates of change.

In Block B you met several mathematical functions and their associated graphs: straight-line (linear) functions and oscillating functions representing musical notes. This block will introduce you to more functions, so that you have a repertoire of mathematical functions to draw upon for different situations. One aim of the block is to build up your library of such functions and techniques for fitting them appropriately to given sets of data. This should enable you to understand the mathematical models used by others as an aid to understanding or prediction, as well as to use mathematics to do some predicting yourself.

Introduction

You have already met many examples of things changing. For example, in *Unit 6* you saw how magnetic north is constantly changing its position, and so the correction to map bearings required to compensate for this alters every year, and in *Units 7* and *8* you met various instances of motion, that is of objects changing their position.

In order to *predict* how things change with passing time, some assumption has to be made about the rate of change. Such an assumption leads to a representation or *model* of the situation.

Often it can be assumed that things go on changing in the same way: that the rate of change is constant. For example, a constant rate of change for magnetic north was assumed in *Unit 6* and constant speeds (rates of change of position) were assumed in *Units 7* and *8*. The assumption of a constant rate of change is useful for an initial simple model as it helps in understanding what is going on while keeping the mathematics reasonably straightforward. If necessary you can improve upon this representation later for greater accuracy of prediction: the accuracy required depends upon your purpose in constructing the model. The assumption of a constant rate of change leads to a straight-line graph called a *linear model*. For short-term predictions linear models may be sufficient, even when they would not be useful longer term.

This unit concentrates on linear models. There are, however, many other sorts of mathematical models, some of which you will see later in this block. Indeed, mathematical modelling is an important activity in many branches of science, social science, technology and other fields where mathematics is applied. The units in this block will thus draw on a wide range of examples in order to illustrate the use of mathematical models in various walks of life. You are expected to understand the modelling contexts from the standpoint of an ordinary adult, not as an expert; and you should try to understand modelling from a commonsense but critical point of view.

When you were working through the first two blocks of the course, you were encouraged to plan and monitor your study, perhaps using the Learning File planning sheets to help you. Clearly, people plan in many different ways and the way you plan is perhaps not as important as actually doing it. It may be useful to look back now to your early planning sheets, from when you started the course. At that stage you may have found studying difficult because you had not had time to develop a system for studying and you were not able easily to identify what you needed to do and when you needed to do it. But by now you may be feeling more confident about studying mathematics and about planning and monitoring your workload. So, to begin this block, spend a few minutes thinking about what has gone well in terms of your planning and about those areas (if any) where you feel you need to make changes.

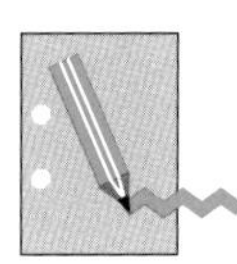

Activity 1 Planning to study Block C

As you begin your study of Block C think carefully about:

◇ how you *manage* your time for study—which involves *finding time* and *using time effectively*;

◇ how you *decide* the work to be done—which involves *allocating time* to the tasks and *monitoring your progress* as you complete them;

◇ your *conditions* of study—the place you study;

◇ how you *organize* all the materials for study.

If there are changes you want to make to the way you study, note these down.

Record your plan for study of this block in your Learning File. Note how you will monitor your progress through the block and how you will assess any changes you make to your study plan.

An important aspect of any plan is that it is a working document. Look at your plan regularly and revise it as you need to.

1 Representing situations involving change

Aims This section aims to introduce you to the modelling process and to the idea of creating a mathematical model for a specific purpose. It revisits some of the standard representations you have met earlier in the course, with the aim of starting a library of standard mathematical models from which to choose in a given situation. ◇

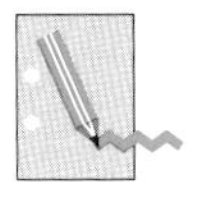

1.1 The process of mathematical modelling

Modelling is perhaps best understood via an example. The example below is quite a simple everyday one, but through it the process of mathematical modelling is illustrated before being generalized and applied in other contexts. An everyday example is used to help you incorporate the concepts within your everyday experience. It should give you a feel for the types of activities which together are labelled 'modelling'.

Example 1 Journey to a tutorial

Imagine that you are one of a group of MU120 students who have sensibly agreed to share a car to travel to a tutorial in Nottingham. Unfortunately, you find yourselves stuck in a stationary queue of traffic on the motorway with about thirty miles still to go. The tutorial starts in three-quarters of an hour. You occupy your time discussing at what time, if you are still stationary, it would become pointless to try to continue on to the tutorial and better to stop at the services, which are one mile ahead, for a self-help group meeting over a cup of coffee instead. The driver estimates that once you get going you could average 60 mph on the remaining 20 miles of motorway, and 30 mph over the final 10 miles of the journey.

The driver's estimates could be called a model. The model is a representation of the situation, designed to help you solve your particular problem.

Using this model for the remainder of the journey, predict how long it will take to reach the tutorial centre once you get going again. After how much longer waiting in the queue of traffic do you think that it would be pointless to continue on to the tutorial?

On the motorway you would travel at 60 mph. That is 60 miles in 60 minutes; so 20 miles would take 20 minutes. After leaving the motorway you would travel at 30 mph. That is 30 miles in 60 minutes; so 10 miles would take 20 minutes. So the total journey time is 40 minutes.

Since there are 45 minutes before the tutorial starts there are 5 minutes to spare. However, the answer you give to the question about when you should abandon your attempts to get to the tutorial depends on how late you are prepared to arrive. If you are prepared to arrive up to 10 minutes late then you would have answered that you would continue on to the tutorial provided that the further delay was no more than 15 minutes. If you felt it was still worth arriving up to 25 minutes late, you would have answered half an hour.

This example indicates that different people may interpret the results obtained from the model in different ways: the interpretation depends on how late you are prepared to arrive at the tutorial.

If the car was stationary for some time, you might use the time to question the original model. Maybe you could average 70 mph on the motorway, but less than 30 mph on the last 10 miles to Nottingham. You might revise the assumptions and repeat the process with different data. Finally, when you started to move again, you would use the results to decide whether to proceed to Nottingham or to stop off at the services for a cup of coffee and a self-help group meeting.

The process you have gone through is called the *modelling process*. It consisted of: identifying the problem (when you should give up trying to get to the tutorial); making some assumptions (about distances and journey speeds) and so creating a model of the journey; using your model (and a little mathematics) to predict the length of time the journey would take; and interpreting the results. Included in the modelling process are any modifications made to the model, together with any repeating of the calculations and reinterpreting of the results.

This process can be stated in a general way, which also applies to other situations, as shown in Figure 1.

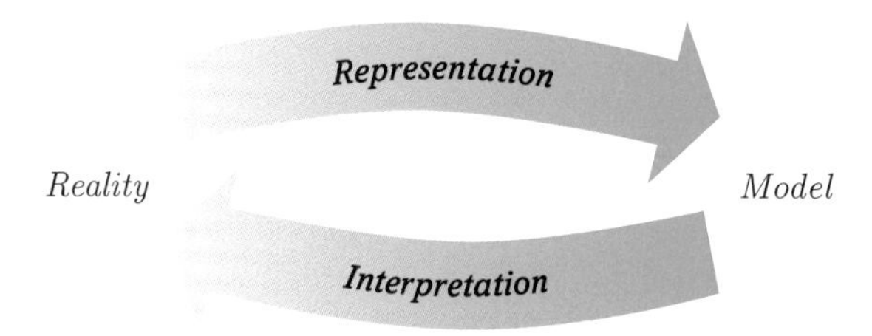

Figure 1 The modelling process

A model is a representation of some aspect of reality created for a specific purpose. If the representation involves mathematics then it is a *mathematical model*. Sometimes the purpose of the model is to solve a problem; or it may be used to make predictions or merely to improve understanding or communicate ideas. However, a model is just a representation—often a simplified representation—of what may be a complex real-life situation, and must be evaluated in that light.

Notice the use of 'a' representation, not 'the' representation, of a real-life situation. There is no *one* 'correct' model for any situation; every situation can be modelled in different ways. The important thing is that the model you choose is appropriate to the purpose for which you wish to use it.

Figure 2 illustrates the modelling process for the simple model described in Example 1.

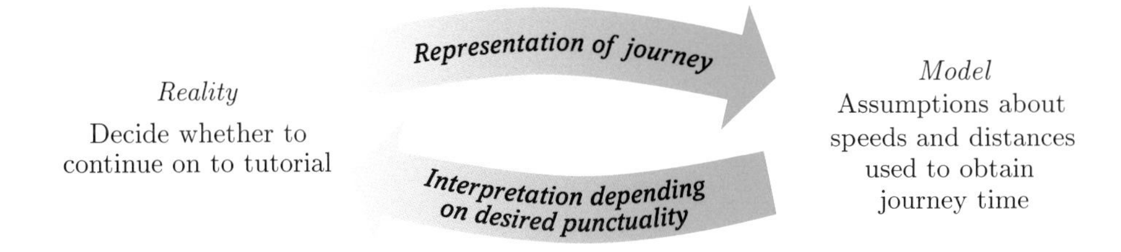

Figure 2

Maps and plans are models. They show a representation of reality by including only features relevant to the purpose. They stress some features and ignore others. Maps, by including different details, are produced for different purposes, as you saw in *Unit 6*. Drawing a sketch map to help a friend find where you live is constructing a model for the purpose of helping your friend. It is a simple representation, not a surveyor's draft, and your friend would interpret it as such. Your map would be a model. As you draw the map, you are making assumptions about what to include and what to leave out. Afterwards, you might discover that it was not quite adequate to the purpose for which it was created—and this is often the case with mathematical models! It may therefore be appropriate to revise it for future use. Figure 3 illustrates the modelling process for this simple model.

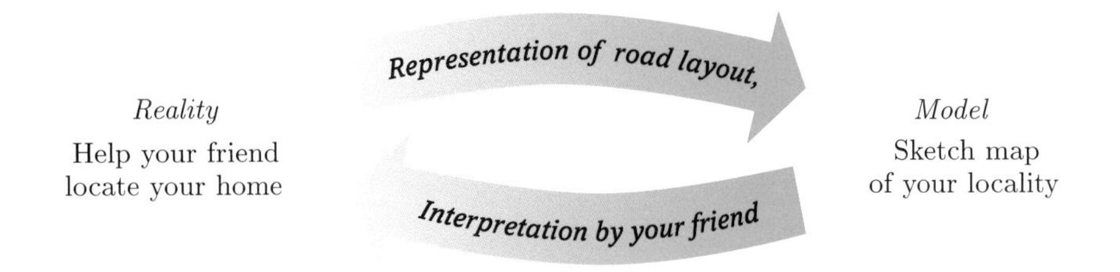

Figure 3

The *modelling process* includes the following stages: first, start with the purpose; second, make assumptions to create a model; then, use the model (perhaps with some mathematics); and, last, interpret the results in the light of the purpose. This process may be repeated several times before the model is considered satisfactory. The process can therefore be a cycle, a *modelling cycle*, which can be repeated as many times as necessary.

Now each stage of *mathematical* modelling will be looked at a bit more closely: first in the context of Example 1; then by generalizing to other situations.

The group of students first discussed the situation and specified the purpose: 'How long should they remain in the stationary queue before it would become pointless to continue on to the tutorial?' The first stage was: *'Specify the purpose'*.

◁ Specify purpose ▷

Next, the driver set up a model of the remainder of the journey. The journey was represented in two parts, motorway and non-motorway, and for each part numerical values (data) were estimated for the distance and speed. The second stage was: *'Create a model'*.

◁ Create model ▷

Then the students did some mathematics to find the time taken for each part of the journey and the total journey time, and obtained a time of 40 minutes. The third stage was: *'Do the mathematics'*.

◁ Do the maths ▷

Then this result was interpreted in the light of the original purpose. The fourth stage was: *'Interpret the results'*.

◁ Interpret results ▷

The student group then questioned the model: was it realistic? was the representation suitable for the purpose? This, additional, fifth stage was: *'Evaluate the model'*.

◁ Evaluate model ▷

Depending upon this evaluation, the students might have gone round the modelling cycle again until they were satisfied or until time ran out. Then the results would be used to determine whether or not to carry on to Nottingham.

These modelling stages can be shown in a modelling cycle diagram, as in Figure 4. This type of formulation of the modelling cycle is useful in many mathematical modelling situations. The broken lines from 'Evaluate the model' indicate that a choice is made here either to repeat the cycle or to use the result. In principle you could go round the cycle as many times as you like, but in practice two or three times are usually sufficient. The first time round the cycle, an oversimplified model may be used to give a feel for the problem and the solution. Then refinements can be added to improve the model until it is considered satisfactory for the purpose for which it has been constructed.

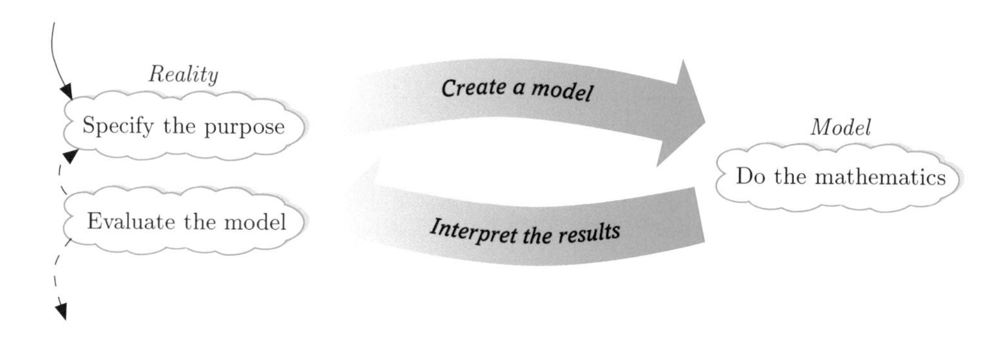

Figure 4 The mathematical modelling cycle

Each stage can help you decide what to do next when modelling a situation yourself or can help you when analysing other people's models. However, you will not find that all mathematical modelling falls neatly into these stages or that it always follows quite such a logical order. Very often two or more stages can overlap, or you may need to backtrack before completing a cycle. For example, you may need to backtrack to remind yourself of why you are doing something; or you may get stuck in trying to find the mathematical solution and so may need to go back and amend the model in order to produce a solvable mathematical problem; or the act of interpretation of results may itself suggest a refinement to the model. The diagram is simply intended to provide a *structural basis* for your modelling, by providing suggestions of the types of activities you should be doing when modelling.

The explicit identification of these different stages of the mathematical modelling process can help you to be systematic in the way you go about modelling and eventually solving your own mathematical problems. For straightforward problems, you may feel you do not need to think about and identify the different stages; but, as you come to attempt more complex ones, breaking the problem down and being systematic in your approach can be extremely helpful.

As you continue to work through this unit, you will often see the different stages identified in the margin. You can adopt this technique with your own problems. You may wish to label the stages as it is done here, or you may wish to create your own shorthand version. Initially labelling may seem awkward and time-consuming, but persevere with its use before you make a final judgement. The important thing, however, is that these stages should begin to become part of your repertoire for solving mathematical problems; they should become so embedded in your thinking that eventually you may not be consciously aware you are solving problems in this way.

To make sure that you can identify the different stages of the mathematical modelling process, consider Example 1 again.

Stage 1: Specify the purpose

◁ *Specify purpose* ▷

The purpose of the model was to answer the question 'How long should the students remain in the stationary queue before it would become pointless to continue on to the tutorial?'

Stage 2: Create a model

◁ *Create model* ▷

Divide the remainder of the journey into two parts and make the following simplifying assumptions about distances and speeds.

1 Motorway: distance 20 miles, average speed 60 mph.

2 Remainder: distance 10 miles, average speed 30 mph.

Stage 3: Do the mathematics

◁ *Do the maths* ▷

Find the time taken for each part of the journey and the total time for the whole journey (using the formula: time = distance/average speed).

Motorway:

$$\text{time} = \frac{20\text{ miles}}{60\text{ mph}} = \frac{20}{60}\text{ hours} = \frac{20}{60} \times 60\text{ minutes} = 20\text{ minutes}$$

Remainder:

$$\text{time} = \frac{10\text{ miles}}{30\text{ mph}} = \frac{10}{30}\text{ hours} = \frac{10}{30} \times 60\text{ minutes} = 20\text{ minutes}$$

The total time for the journey is therefore 20 minutes + 20 minutes, which is 40 minutes.

Stage 4: Interpret the results

◁ *Interpret results* ▷

From the time that they emerge from the queue, it will take them about 40 minutes to reach the tutorial centre, leaving 5 minutes to spare. Hence it is worth going on if they are not stuck for more than about another 15 minutes (allowing for being 10 minutes late).

Stage 5: Evaluate the model

◁ *Evaluate model* ▷

The students might discuss how realistic the original model was: were the distances and speeds about right? could they arrive only up to 10 minutes late? If they were dissatisfied with the model, they would change it and go round the modelling cycle again. If they were happy with the solution, they would move on to devise a plan of action based on this result: for example, 'If we are stuck for another 15 minutes or less we go on to the tutorial. Otherwise we go to the services and have a self-help group meeting and a drink.'

Different types of modelling may have different associated modelling stages, and even within mathematics there is much discussion about the naming of the stages and even about the exact number of stages. Hence the diagram and the five stages are by no means prescriptive. They are a useful guide to help you in using mathematical modelling for different purposes.

To illustrate this point, consider the statistical investigation cycle from *Unit 5*, shown in Figure 5. Statistical investigation is a modelling activity, and so the statistical investigation cycle is a modelling cycle, which is different from but yet has many similarities with the cycle in Figure 4.

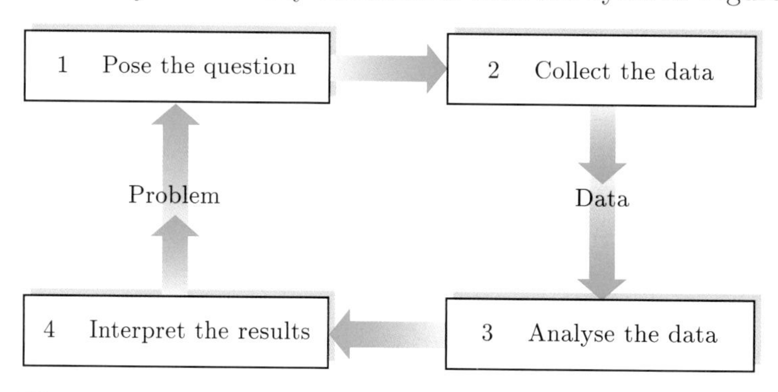

Figure 5 The statistical investigation cycle

Activity 2 What is mathematical modelling?

Suppose that a friend had glanced over your shoulder as you read this subsection and expressed an interest in the headings and margin notes. Write down a few notes on how you would explain mathematical modelling and the stages in the modelling cycle to such an interested person.

1.2 Choices of representation

When you have a situation to model mathematically, you will usually have a choice of representations. For instance, in Block A (*Unit 5*) the number of seabirds counted at different sites was taken as a representation of the total population of birds and it was implicitly assumed that the way the data varied represented similar variations in the total population. In fact the data were collected on several occasions, often by several people, and an average was taken as the representation. There was a choice as to which average: either the mean or the median could have been used.

In Block B you saw that there is often a choice between a graphical representation and an algebraic representation. The choice depends not only on the situation being represented but also on the purpose of the representation. In this block you will be using both graphical and algebraic representations of different situations for a variety of purposes.

There are a number of standard representations: a library of standard representations from which to choose in any given situation. You already have some representations in your library:

- ◇ *statistical representations*: scatterplots, boxplots, medians and interquartile ranges, means and standard deviations;
- ◇ *graphical representations*: straight-line conversion graphs, distance–time graphs;
- ◇ *algebraic representations*: formulas and equations, mathematical functions like x, x^2, $\sqrt{x}$, $\sin x$.

By the end of the block, your library of standard representations will be considerably larger.

The advantage of a library of representations is that much of the mathematics of each representation in the library is well known. You do not have to reinvent the wheel in every model. You can learn about the different types of wheel (representations) and then use whichever is the most suitable one for your purpose in any particular case.

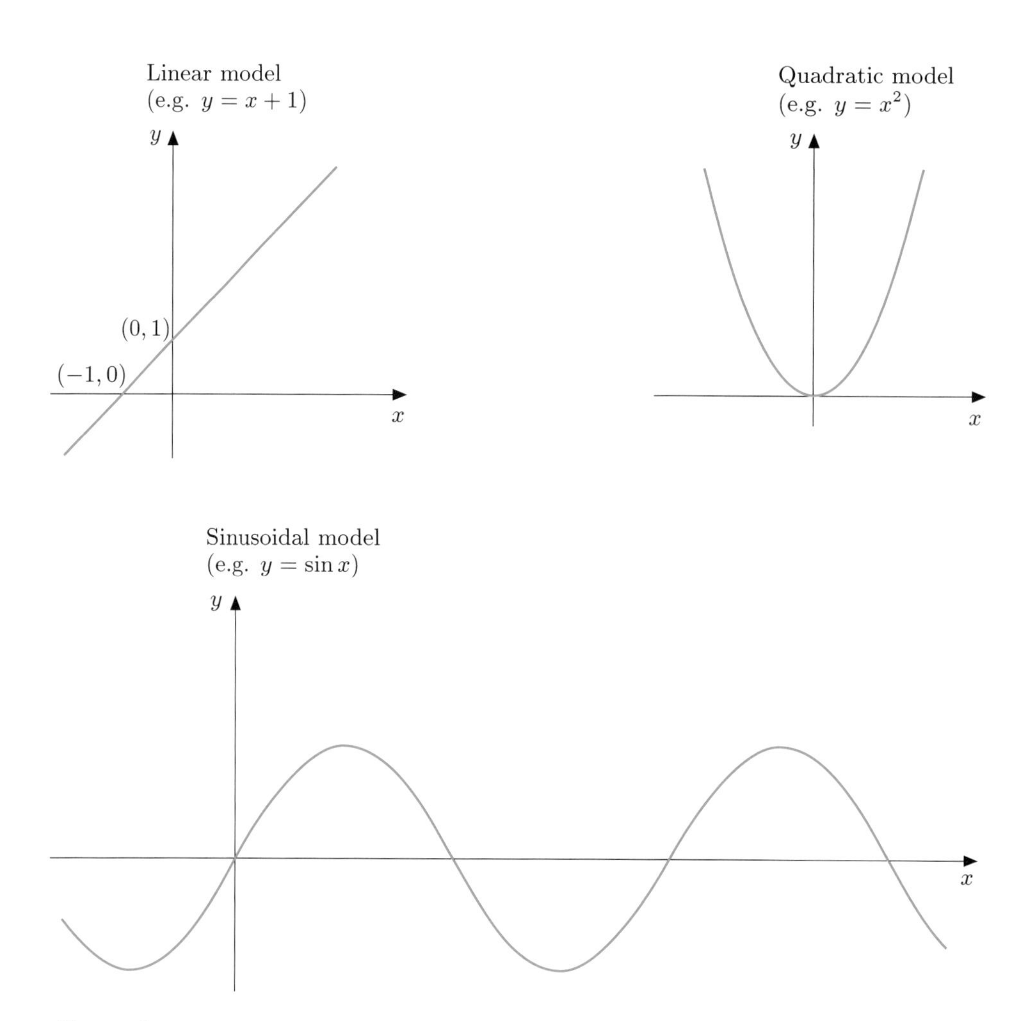

Figure 6

Figure 6 shows three standard graphical models, based on functions you have already encountered (which have both graphical and algebraic forms). Each model is based on a different assumption. For example, a *linear* (straight-line) model assumes a constant rate of change (or gradient); a *sinusoidal* model assumes repetition. Considering which assumption is appropriate in different situations—that is, which representation suits your purpose best—is an important modelling skill. Sometimes you have data upon which to base your choice of representation: for example, from a survey or from background information or theory.

Activity 3 Choosing a representation from your library

Look at each of the data sets represented graphically in Figure 7 and decide which of the models from Figure 6 would be most appropriate to represent the data over the range of values covered.

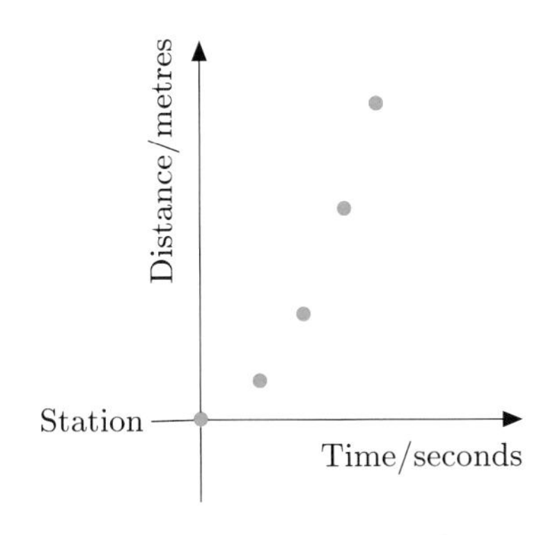

(a) Distance–time graph of train leaving station

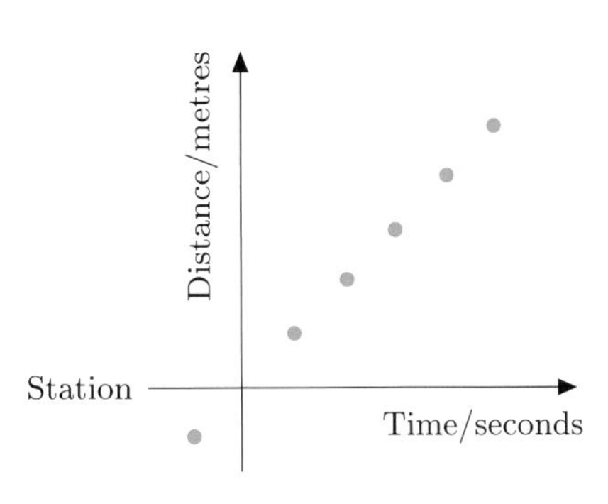

(b) Distance–time graph of express train going through station

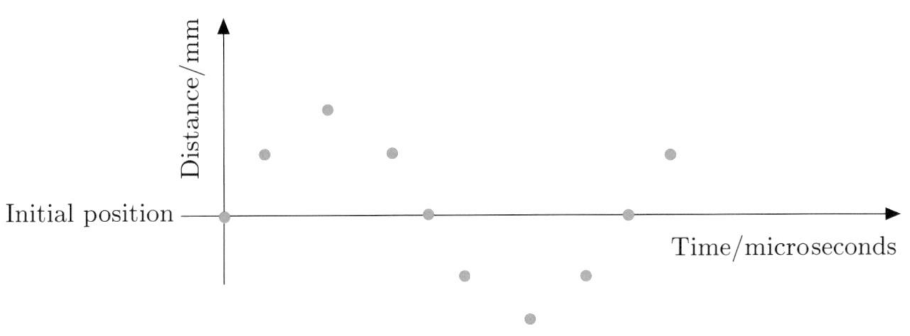

(c) Distance–time graph of tip of tuning fork

Figure 7

This unit concentrates on linear models. These can be used in a huge variety of situations. Also, because their mathematics is straightforward, they are often preferred to more complicated models in the first stages of the modelling process or when the range of values for which the model is required is small.

Activity 4 Models appropriate for small ranges

Consider the following three algebraic models:

$y = x$ (linear)

$y = x^2$ (quadratic)

$y = \sin x$ (sinusoidal)

(a) Enter these functions into your calculator (make sure the calculator is in radian mode) and plot the graph of each function, in turn, for the range of values of x from -3 to $+3$ and y from -9 to $+9$. (You might like to refer back to Chapters 8 and 9 of the *Calculator Book* for instructions on drawing graphs of functions on your calculator.)

Describe the shape of each graph briefly, as if to a friend who knows little mathematics. (Try saying your description out loud.)

(b) In science the function $y = \sin x$ is sometimes approximated by $y = x$. From your graphs, suggest a region where the two graphs are very close together, and therefore where this approximation might be useful.

(c) Repeat part (a) but with the range of values of x from 0.1 to 0.2. Adjust the range of y until you can see the whole graph in each case. Remember to describe the shape of each graph briefly, as if to the same friend (and again out loud).

So, as Activity 4 has illustrated, a linear model can often be a useful one for modelling over a small range. However, care needs to be taken if the range is extended.

◁ *Create model* ▷

One important choice of representation that often has to be made is whether to choose a graphical or an algebraic model. Often a graphical model is sufficient, but frequently the additional flexibility provided by an algebraic model is preferable. To illustrate this, consider again the journey of Example 1. The arithmetic model used there could have been represented graphically by a distance–time graph, as shown in Figure 8.

Remember that constant speed leads to a straight-line distance–time graph.

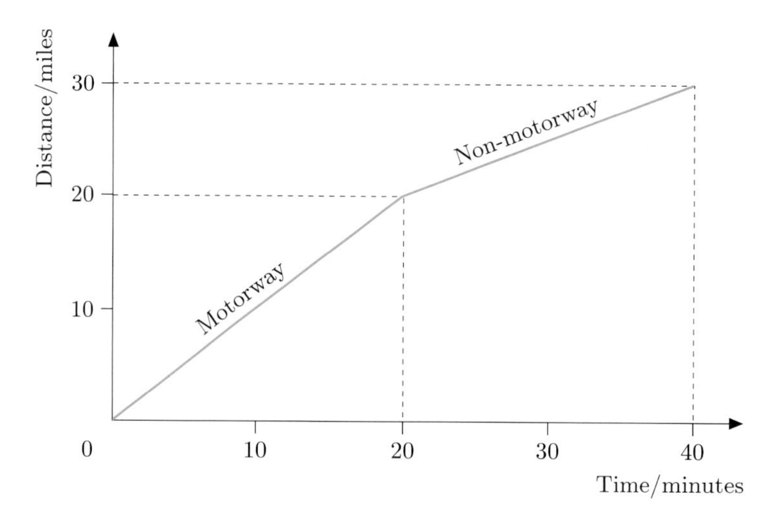

Figure 8

The gradient of each line represents the speed—the steeper line representing the motorway part of the journey (1 mile per minute) and the less steep line the non-motorway part of the journey ($\frac{1}{2}$ mile per minute).

This representation is fine for use with the assumptions made in Example 1, and it gives a good feel for what happens at different speeds. But what if the students wanted to examine the effect of different assumptions? Varying the assumptions about the speeds would alter the steepness of the lines, and a different graphical representation would need to be drawn. This could be quite time-consuming. An algebraic representation, however, would be quicker to modify for different modelling assumptions about speeds.

Suppose the speed on the motorway is S_1 miles per hour and that on the non-motorway roads is S_2 miles per hour. Then:

◁ *Create model* ▷

$$\text{motorway time} = \frac{\text{distance}}{\text{speed}} = \frac{20}{S_1} \text{ hours;}$$

S_1 is read as 'S one'.

$$\text{non-motorway time} = \frac{10}{S_2} \text{ hours;}$$

$$\text{total time} = \frac{20}{S_1} + \frac{10}{S_2} \text{ hours.}$$

This might be more conveniently expressed in minutes and so could be rewritten

$$\text{total time} = \left(\frac{20}{S_1} + \frac{10}{S_2}\right) \times 60 \text{ minutes.} \qquad (1)$$

Now the original assumption was $S_1 = 60$ and $S_2 = 30$. Substituting these values in equation (1) gives

◁ *Do the maths* ▷

$$\text{total time} = \left(\frac{20}{60} + \frac{10}{30}\right) \times 60 = 40 \text{ minutes.}$$

However, if the alternative optimistic assumption of $S_1 = 70$ and $S_2 = 50$ is made, equation (1) gives

$$\text{total time} = \left(\frac{20}{70} + \frac{10}{50}\right) \times 60 = 29 \text{ minutes (to the nearest minute).}$$

Other values for the two speeds S_1 and S_2 could be substituted in equation (1) and so a feeling for the range of likely journey times obtained.

Activity 5 Using the algebraic model

(a) If the driver were more cautious he might suggest speeds of 50 mph on the motorway and 25 mph for the remainder of the journey. Use equation (1) to calculate the total journey time with these assumptions.

(b) Assume your answer to (a) is the most pessimistic estimate and the 29 minutes above the most optimistic estimate. Describe, as if to a fellow student in the car, how the algebraic and graphical models have aided your understanding of the situation.

◁ *Interpret results* ▷

◁ *Evaluate model* ▷

To summarize, the graphical model is fine under the assumptions of Example 1 and gives a good picture of what happens at different speeds. However, it is not so useful for considering alternative assumptions about the speeds, as it would be time-consuming to have to draw different graphs for each alternative assumption. The algebraic model enabled alternative speeds to be substituted in the equation easily in order to try out different assumptions: for example, it was easy to compare the most optimistic and the most pessimistic estimates.

In general, a graphical model shows the relationship between two variables clearly, but it takes time to draw and it is less flexible and can be less accurate than an algebraic model. Often a graphical model is to be preferred, but more often an algebraic model is better (and sometimes even a combination of the two can be most useful).

Do not worry if at the moment you find algebraic symbols and equations rather abstract—you will gradually become more fluent in their use and more appreciative of the power of algebraic models. Until your fluency develops, you can always keep reminding yourself of what the symbols stand for, or you can use your calculator to draw graphs of any algebraic equations or formulas.

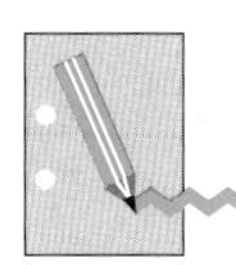

Activity 6 Learning mathematics through reading

The Learning File activity that runs through this unit asks you to think about how you learn mathematics through reading. To begin, think about how you 'read' the text. What have you just been reading in this section? Were you able to identify the underlying ideas through the section? Did any of these ideas relate to anything you already knew? Have you been able to pick out the essential points, and if so how did you identify them?

When you are satisfied with your response to these questions, take a few minutes to think about how you have been using the text so far on this course, and try to summarize your thoughts.

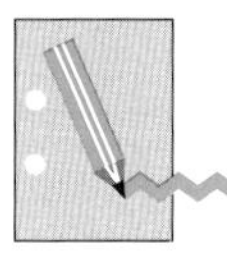

Activity 7 Handbook work

There are two Handbook sheets associated with this unit: one is for recording definitions or explanations of new terms and the other is for describing modelling techniques. Look back through this section and bring these sheets up to date. (You should continue to do this throughout the unit.)

In this section, you have seen that a *model* is a representation of reality created for a specific *purpose* and that modelling including mathematics is called *mathematical modelling*. It is sometimes useful to think of the mathematical *modelling process* in terms of the following sequence of activities:

Specify the purpose
Create a model
Do the mathematics
Interpret the results
Evaluate the model

If appropriate, repeat the *modelling cycle*, revising the model, until you are satisfied with the revised model. Modelling does not always follow such a logical sequence: the stages listed above provide suggestions, rather than prescriptions.

There is often a choice of mathematical representation: for example, linear, quadratic or sinusoidal in either graphical or algebraic form. You should build up a library of standard representations to use in different situations.

Outcomes

After studying this section, you should be able to:

◇ describe to someone else the type of activities involved in mathematical modelling (Activity 2);

◇ describe and compare the graphical and algebraic models of the type mentioned in this section (Activities 3, 4, 5).

2 Linear models

Aims This section looks at situations where a constant rate of change is assumed, leading to a straight-line graph (sometimes known as a linear model). It aims to help you revise graphical and algebraic representations and to help you create linear models and use them to make predictions. It also aims to revise how the gradient and intercept of a graphical linear model are related to the corresponding algebraic model. ◇

2.1 Graphical linear models

Creating a mathematical model or representation (stage 2 of the modelling cycle) is not always easy to do. It involves translating words into mathematics. Experience will help you, and the more practice you have the easier you will find it. First identify the relevant variables and then make suitable assumptions. Recall that *variables* are quantities which are not fixed; they may change (like time or the distance you have travelled). Then use these variables and assumptions to create the model.

This section deals with one of the simplest forms of model, a *linear model*, where the relationship between the variables gives a straight-line graph, a *linear graph*. Linear models are often very useful, especially for short-term predictions.

Look now at an example of the sorts of assumptions that lead to a linear model.

Example 2 Motorway journey

◁ *Specify purpose* ▷

In the course of my work, I often have to travel from Nottingham to Milton Keynes, a distance of 87 miles (almost all on the M1 motorway), which usually takes $1\frac{1}{2}$ hours. I often give people lifts from Nottingham, but on this particular occasion I have agreed to pick up someone on the M1 at a junction near Leicester, 25 miles from Nottingham. When am I likely to get to the junction if I leave Nottingham at 8 am?

◁ *Create model* ▷

First, what are the relevant variables? They are the distance travelled and the time taken. These are the two important quantities in the situation.

The main assumption is about speed. Since most of the journey is along the motorway, it is reasonable and straightforward to assume a constant speed for the whole journey from Nottingham to Milton Keynes.

◁ *Do the maths* ▷

So, under the assumption of constant speed, half the journey ($43\frac{1}{2}$ miles) takes half the time (0.75 hour), a third of the journey (29 miles) takes a third of the time (0.5 hours), and so on, as in Table 1.

Table 1

Time taken (hours)	0	0.5	0.75	1	1.5
Distance travelled (miles)	0	29	43.5	58	87

Plotting these points by hand on a distance–time graph, in order to predict the time for the 25 miles to the Leicester junction, gives the diagram shown in Figure 9.

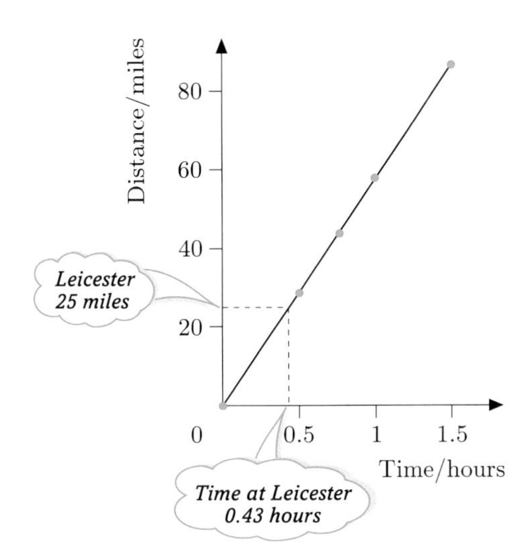

Figure 9

The points lie on a straight line. This is a direct consequence of the assumption of constant speed. Reading off the time for a journey of distance 25 miles gives 0.43 hours. So I should reach the Leicester junction after about 0.43 hours (26 minutes).

◁ *Interpret results* ▷

However, you should be aware of the limitations of the model: it is only valid for a motorway journey of duration up to $1\frac{1}{2}$ hours and has not taken into account variations in speed. So I am unlikely to arrive after exactly 26 minutes. I would tell my passenger to expect me between 8.20 and 8.30 am.

Notice that because a constant rate of change (in this case, speed) was assumed, two points would have been sufficient in order to draw the straight line in Figure 9.

In Example 2, the assumption of constant speed leads directly to a straight-line graph (a line of constant slope). The *gradient* (or slope) of the line gives the assumed constant speed:

$$\text{speed} = \frac{\text{distance}}{\text{time}} = \frac{87\text{ miles}}{1.5\text{ hours}} = 58\,\text{mph}$$

$$\text{gradient} = \frac{\text{increase in distance}}{\text{increase in time}} = \frac{87\text{ miles}}{1.5\text{ hours}} = 58\,\text{mph}$$

Notice from Table 1 that 1 hour is taken to travel 58 miles, which confirms this speed.

2.2 Algebraic linear models

Much of what has been covered using words and graphs can equally well be expressed algebraically using symbols and equations. An algebraic equation shows the relationship between variables, just like a graph. Instead of a graph of 'distance travelled' against 'time taken', for example, there will be an equation or formula relating d (representing 'distance travelled') and t (representing 'time taken'). The equation which relates d and t will be true for *all* the points on the graph and it describes uniquely both the graph and the relationship between the variables. It is possible to translate from an equation to its graph and vice versa: they are equivalent. However, the equation can be written down quickly and in a fraction of the space needed for its equivalent graph.

You could have equally well chosen symbols other than d and t, but it is often helpful to choose symbols that help you remember what they represent. However, once you have chosen the symbols, stick to them.

Consider the motorway journey in Example 2, using symbols. Let d miles be the distance travelled by the car in a time t hours. You saw above that the gradient of the graph and the constant speed are the same: 58 mph. The speed is distance travelled (d miles) divided by time taken (t hours), so:

$$\frac{d}{t} = 58$$

This is an algebraic equation showing the relationship between the variables. However, it is more useful to rearrange the equation so that just one of the variables is on the left-hand side of the = sign. If two expressions like '$\frac{d}{t}$' and '58' are equal, then multiplying both by the same number again produces two equal expressions. So multiplying both sides of the equation by t gives:

$$d = 58t$$

This equation also relates the variables d and t, but from this it is easier to determine the distance travelled at any given time.

You should always write down the limitations of an algebraic model. In this case, the equation is only valid for the duration of the journey described in the example. In other words, the variable t can only take values between 0 and 1.5. This can be written neatly as:

$$0 \leq t \leq 1.5$$

If you have not met this type of inequality sign before or are unsure of its meaning then read the following box on inequality signs.

Inequality signs

Inequality signs are used as a shorthand notation. They supplement the equals sign. Below is each sign and its meaning in words, together with an example.

Symbol	Meaning	Example
$=$	is equal to	$t = 3$ means 't is equal to 3'.
$<$	is less than	$u < 3$ means 'u is less than 3'. So u can be any number less than 3.
$\leq$	is less than or equal to	$x \leq 3$ means 'x is less than or equal to 3'. So x can be any number less than 3, or 3 itself.
$>$	is greater than	$y > 3$ means 'y is greater than 3'. So y can be any number greater than 3.
$\geq$	is greater than or equal to	$v \geq 3$ means 'v is greater than or equal to 3'. So v can be any number greater than 3, or 3 itself.

The smaller end of the sign is next to the smaller number, and the larger end of the sign is next to the larger number. For example:

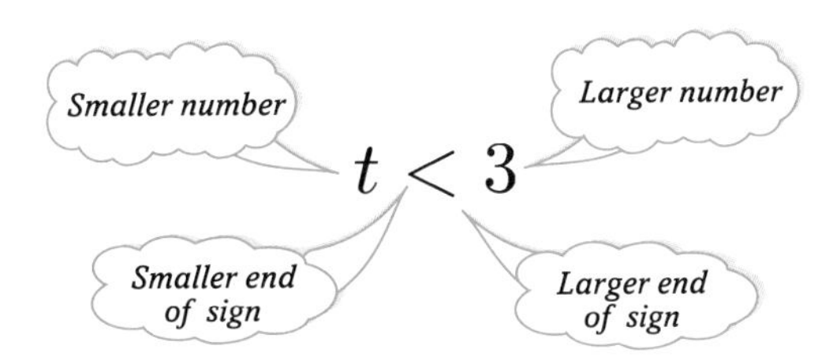

So t is less than 3; that is, t is smaller than 3.

The signs are sometimes used in pairs to indicate that a quantity lies between two numbers. So, for example,

$$3 < t < 5$$

means '3 is less than t and t is less than 5' or 't lies between 3 and 5'.

The equation $d = 58t$ is valid for every value of t from 0 and 1.5 and so can be used to find out how far the car will have gone after any particular time between 0 and 1.5 hours from the start of the journey, say half an hour.

$$t = 0.5$$
$$d = 58t = 58 \times 0.5 = 29$$

So the car will have gone 29 miles in half an hour.

You can also use the equation to find the time at which the car will be at any particular place. The problem was 'How long will it take me to travel the 25 miles to the Leicester junction?' or 'What is t when $d = 25$?' Substituting $d = 25$ into the equation gives:

$$25 = 58t$$

Dividing through by 58 gives:

$$t = \frac{25}{58} = 0.431\,034\,48 = 0.43 \text{ (rounded to two decimal places)}$$

This is the same value as found from the distance–time graph: 0.43 hours or 26 minutes.

The equation

$$d = 58t \qquad (0 \leq t \leq 1.5)$$

is called the *equation of the graph*. The equation of a linear graph is often called a *linear equation*. It gives the relationship between the variables in the linear model, which is called a *linear* relationship.

Now, try to derive and use such an algebraic equation yourself, to model the situation described in Activity 8.

Activity 8 Cross-channel ferry

A cross-channel ferry normally takes 2 hours to make the 40 km crossing from England to France.

(a) Create a simple algebraic linear model of the ferry's motion, which might be used—when travelling on the ferry—to tell you how far you have travelled, by the following stages.

(i) Determine the relevant variables and allocate symbols to represent them.

(ii) Assuming that the speed is constant, find the average speed of the ferry.

(iii) Find the algebraic equation that relates the relevant variables.

(iv) State the limitations of the model.

(b) From your model find:

(i) how far the boat has travelled after $\frac{1}{4}$ hour, and after 1 hour 10 minutes;

(ii) after how long the boat is 15 km from the English port, and 35 km.

There is quite a bit of vocabulary concerning algebraic equations, some of which you have met before. When d, the distance travelled in miles, depends upon t, the time taken in hours, then d is called the *dependent variable* and t is called the *independent variable*. The independent variable is plotted on the horizontal axis and the dependent variable on the vertical axis. Because d is dependent upon t it is often said to be a *function* of t. Distance travelled is a function of time. (The dependent variable is a function of the independent one.)

It is not always clear which variable should be the dependent one and which the independent one, and as you will see shortly it does not always matter. (For instance, one might have formulated the motorway journey example in terms of how long it would take to travel a known distance d. Then the answer is given by $t = d/58 : t$ as a function of d.) Time is often the independent variable. In experimental work, the independent variable is the one the experimenter can control and the dependent variable is the one that must be measured.

The equation $d = 58t$ can be thought of as specifying a function: any value of t can be substituted into the equation to get the corresponding value of d. When d is a function of t the notation $d = f(t)$ is sometimes used (read as 'd equals f of t' or 'd is a function of t'). In the motorway journey example, the function $f(t)$ is $58t$.

A *function* may be thought of as a process which converts one number (the *input* value) into another (the *output* value).

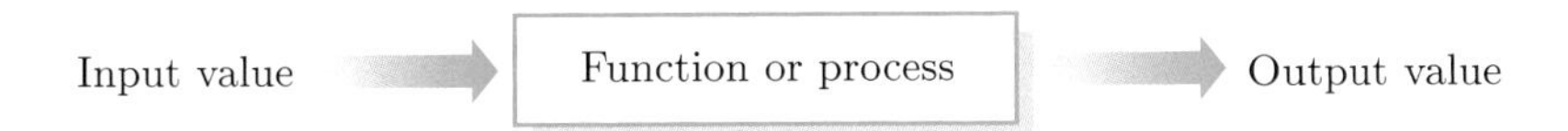

In the motorway journey example, the input value is t (the time in hours), and the output value is d (the distance in miles), which is 58 times the input value t.

t → Multiply by 58 → $d = 58t$

For every input value there must be one and only one output value.

You can use your calculator (or a computer) to perform the process. Take the function $d = 58t$. You input the value of t into your calculator and press × **5 8** ENTER, and then you have the corresponding value of d in the display of your calculator.

Many other functions can also be represented by a sequence of keys to press on your calculator.

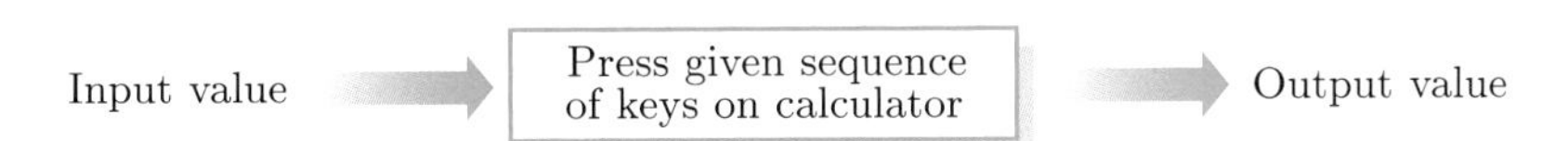

There is one additional piece of information which should be included in the specification of a function, namely any limitation on the input value. In the case of the function $d = 58t$, you need to specify $0 \leq t \leq 1.5$, because t is limited to the numbers between 0 and 1.5 (inclusive). This is sometimes referred to as the *range of the variable t* or the *domain of the function.*

You may also hear the phrase 'the graph of the function', which means exactly what it says. The graph of the function $d = 58t$ $(0 \leq t \leq 1.5)$ is the straight-line graph shown in Figure 10.

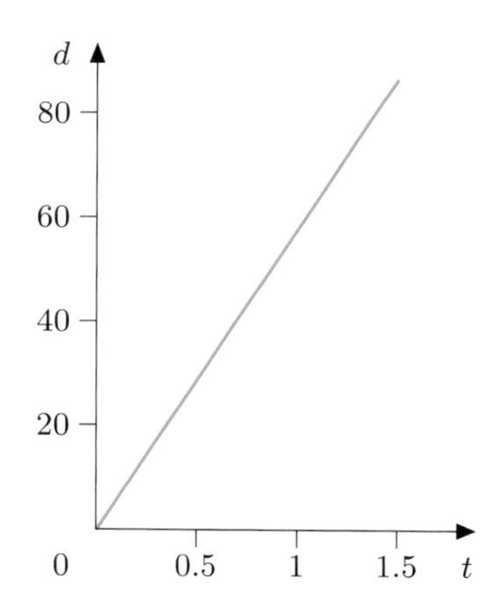

Figure 10

The equation $d = 58t$ $(0 \leq t \leq 1.5)$ can be rearranged as

Remember that the total distance travelled on the whole journey is 87 miles.

$$t = \frac{d}{58} \qquad (0 \leq d \leq 87),$$

in which case d (the distance travelled) is the independent variable and t (the time taken) is the dependent variable. In this case t is a function of d: this is written $t = f(d)$. The graph of this equivalent function would be like that in Figure 10 but with the axes reversed. This is called the *inverse function* of $d = 58t$.

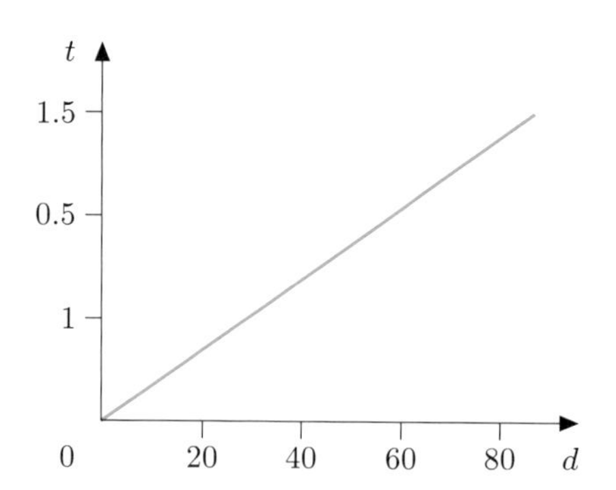

Figure 11

Speed is the rate at which distance travelled increases with time, and a constant speed implies that the graph of distance against time is a straight line. The same principle applies to other variables increasing at a constant rate: the corresponding graphs are straight lines.

Activity 9 Snow-ploughs

The roads in a country district were completely free from snow when it started snowing at midnight one winter's night, and it snowed steadily thereafter. At 10 am, when the snow was 19 cm deep, the snow-plough supervisor had to make a decision about what time to tell the snow-plough drivers to start work. Ideally the snow should not be more than 30 cm deep on a road when the snow-ploughs start to clear it. However, the district council is trying to economize and so the snow-ploughs should not be called out too early.

(a) What are the relevant variables?

(b) What straightforward assumption needs to be made?

(c) Draw a graph to model the situation, and hence predict when the snow will be 30 cm deep.

(d) If you were the supervisor, what would you decide?

The snow-plough problem has a constant rate of increase of the depth of snow of 1.9 cm per hour, which corresponds to the gradient of the graph shown in Figure 44.

The rate at which the depth (d cm) of snow is increasing is 1.9 cm per hour. In t hours there will be $1.9t$ cm of snow, since d is 0 initially. So the equation relating d and t is:

$$d = 1.9t$$

This equation is a valid model for $t \geq 0$ and until it stops snowing. You can use the equation in order to solve the problem in Activity 9: 'When will the snow reach a depth of 30 cm?' or 'When $d = 30$, what is t?' Substitute $d = 30$ in the equation to give:

$$30 = 1.9t$$

So $t = \frac{30}{1.9} \simeq 15.79$.

So a depth of 30 cm is reached at about $15\frac{3}{4}$ hours after it started to snow: that is, at about a quarter to four in the afternoon, as found previously.

The linear graphs for the motorway journey and snow-plough examples both passed through the *origin* (the point $(0, 0)$ where the axes cross). When the graph does not pass through the origin, you have to be a little more careful in determining the corresponding algebraic equation.

Example 3 Flood warning

◁ *Specify purpose* ▷

A major river is rising fast and the authorities are worried about the possibility of flooding. Several readings of depth have been taken, as shown in Table 2. The banks of the river are 26 metres high. By when should the authorities have evacuated people living on low-lying ground near the river?

Table 2

Time	midnight	2 am	4 am	6 am
Depth (metres)	5	9	13	17

◁ *Create model* ▷

The relevant variables are the depth of the river in metres (say, d) and the time in hours after midnight (say, t). Since the river is rising steadily, it is reasonable to assume that the water level will continue to rise at the same constant rate until the river breaks its banks. Plotting d against t gives a straight-line graph (see Figure 12).

The time variable could be chosen differently, for example t hours after 6 am; the choice of actual time when $t = 0$ is quite arbitrary. This choice was based on ease of finding t from a given time: on this scale when $t = 6$ it is 6 am.

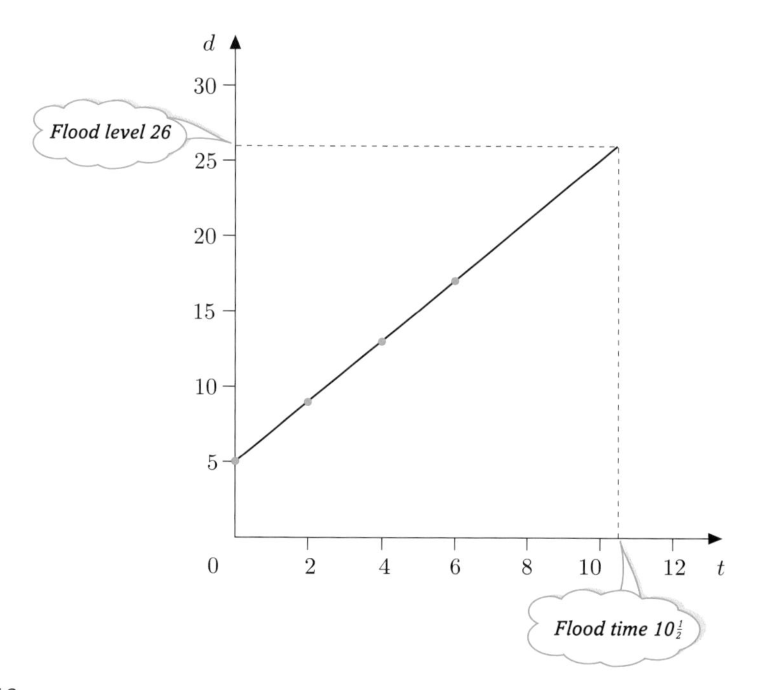

Figure 12

◁ *Do the maths* ▷

The assumption of a linear relationship between depth and time (between d and t) is valid from $t = 0$ up until $d = 26$ (the height of the banks). When $d = 26$ (when the river will flood), the time is $t = 10\frac{1}{2}$ or 10.30 am.

◁ *Interpret results* ▷

The model predicts that the river will flood at 10.30 am. But the authorities would be wise to allow a good safety margin in case the river starts to rise faster than predicted. They would probably aim to clear the vulnerable areas well before 10 am, and to keep monitoring the depth carefully up until then, modifying their model if the rate at which the water was rising changed significantly.

This example assumed that the river would rise at a constant rate, giving a linear graph of d against t; but the line does not pass through the origin. The equation of such a line is obtained in a similar way to when the graph passes through the origin, but an allowance must be made for the initial value—where the graph cuts the vertical axis.

In the flood-warning example, the rate of increase in the depth can be calculated from the data in Table 2. It is:

$$\text{rate of increase in depth} = \frac{\text{increase in depth}}{\text{increase in time}} = \frac{(17-5)\text{ metres}}{(6-0)\text{ hours}}$$
$$= 2 \text{ metres per hour}$$

So the gradient or slope of the graph of d against t is 2.

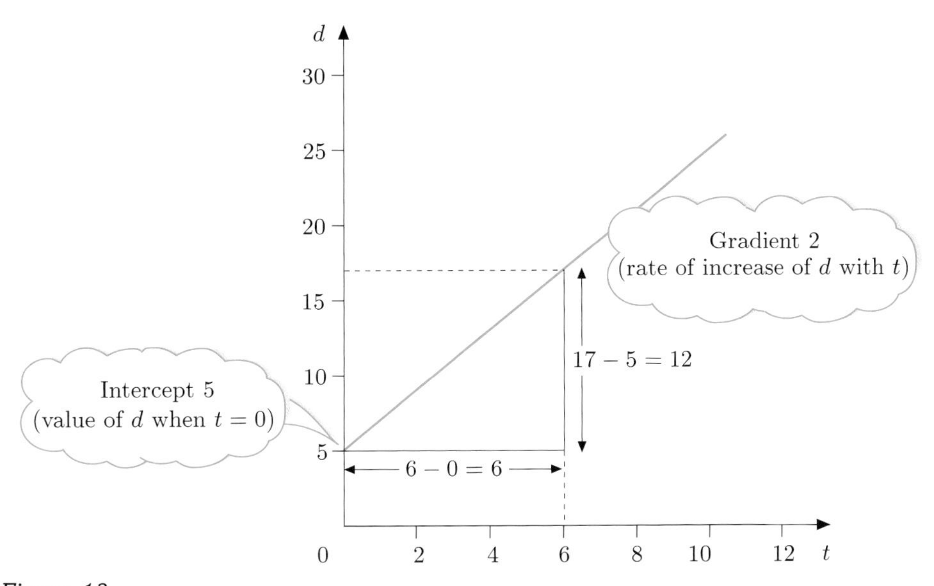

Figure 13

The graph cuts the vertical axis when d is 5. Recall that this is called the *intercept*: $d = 5$ (a depth of 5 metres) when $t = 0$. The depth at subsequent times will be 5 metres plus the increase since $t = 0$ (midnight). Since the depth is increasing at 2 metres per hour, after t hours d will have increased by $2t$. This must be added on to the initial depth (5 metres); so, at time t, $d = 2t + 5$.

This model is valid for $t = 0$ and up until the river bursts its banks, so it can be written down as follows:

$$d = 2t + 5 \qquad (t \geq 0, d \leq 26)$$

Notice the form of this equation:

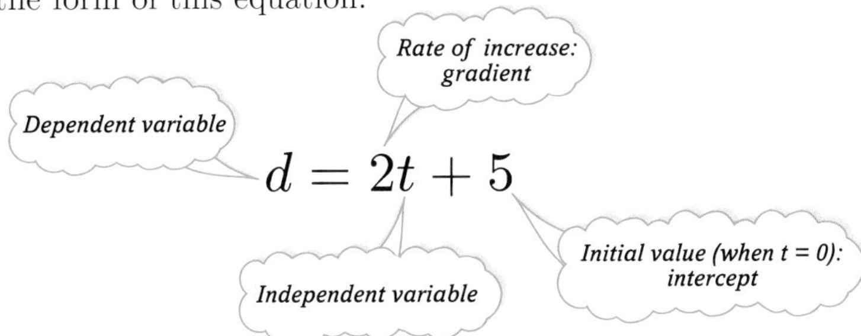

2.3 The equation of a straight line

The equation of any straight line is of the form:

$$\text{dependent variable} = \text{gradient} \times \text{independent variable} + \text{intercept}$$

The equations of *all* linear graphs can be written in this form and all equations of this form are linear equations (representing linear graphs). You can write down the equation of a straight-line graph by finding the *gradient* (slope) and the *intercept* (the value of the dependent variable when the independent variable is zero; that is, where the graph crosses the vertical axis). You may be familiar with the form of the equation of a straight line as

$$y = mx + c$$

where y is the dependent variable, m the gradient, x the independent variable and c the intercept. In this formulation m and c stand for numbers. They are *constants*, whereas x and y are *variables*.

Activity 10

Write down the gradient, the intercept and the equation of each of the straight lines in Figure 14.

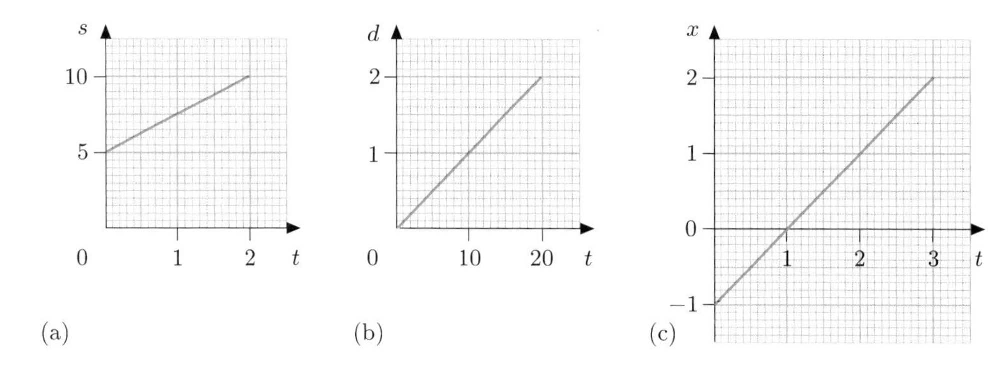

Figure 14

The format of the equation of a straight line (linear graph) is the same irrespective of whether the intercept or the gradient is positive or negative:

$$\text{gradient of straight line} = \frac{\text{increase in dependent variable}}{\text{increase in independent variable}}$$

A decrease in either variable is interpreted as a negative increase.

Example 4 *Finding linear equations*

Find the equations of the straight lines shown in Figure 15.

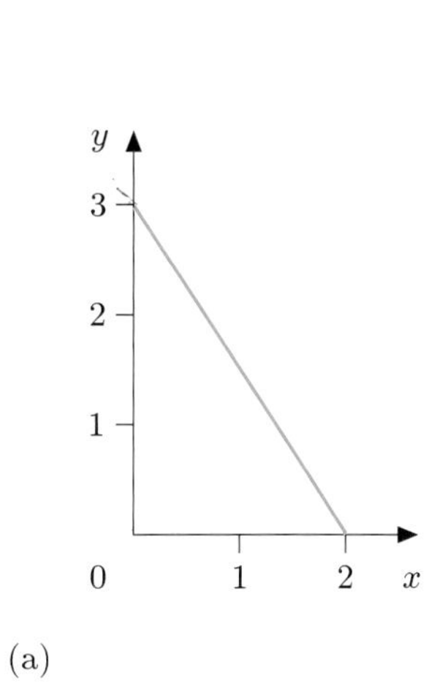

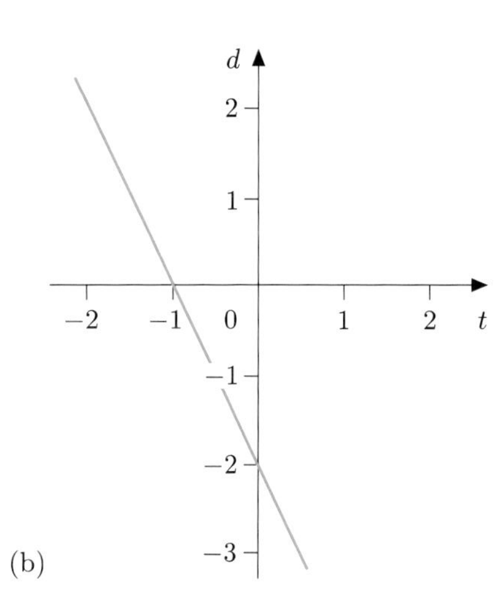

Figure 15

(a) The intercept is 3. The gradient is:

$$\frac{0-3}{2-0} = -\frac{3}{2}$$

So the equation is:

$$y = -\tfrac{3}{2}x + 3$$

(This might be rewritten without fractions by multiplying through by 2. This gives:

$$2y = -3x + 6$$

An equivalent form is:

$$2y = 6 - 3x)$$

(b) The intercept is −2. The gradient is:

$$\frac{-2-0}{0-(-1)} = \frac{-2}{1} = -2$$

The equation is:

$$d = -2t - 2$$

Activity 11

Find the equation of each of the straight lines in Figure 16.

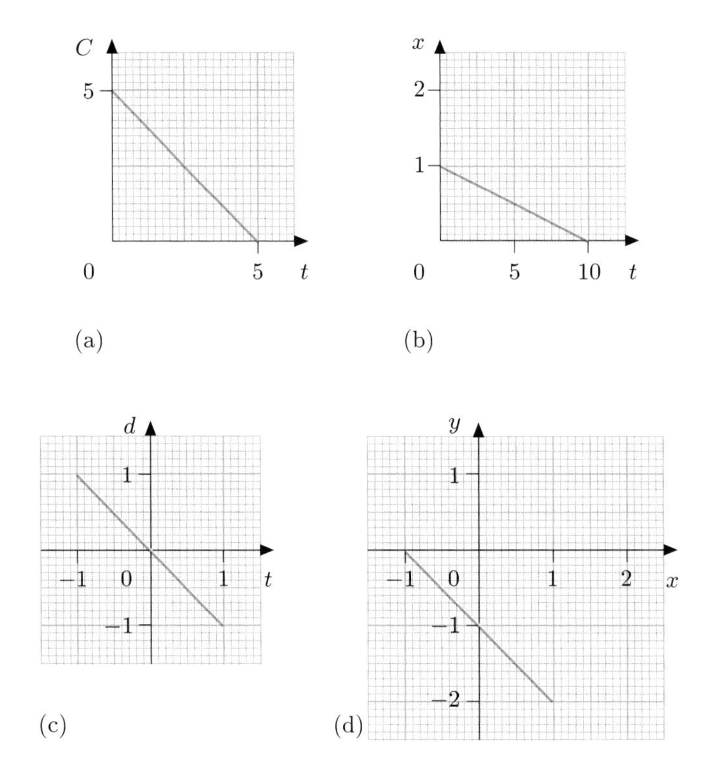

Figure 16

There is often more than one way to formulate a linear model. For instance, in Example 3, the dependent variable could have been 'height of bank above river', denoted by h and measured in metres. Then $h = 26 - d$.

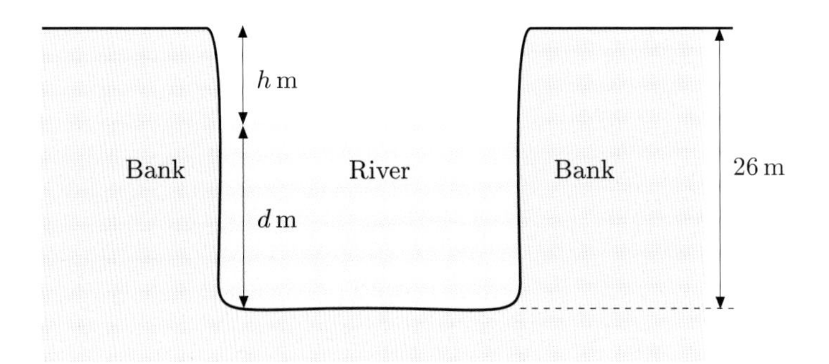

Figure 17

A model could then be formulated to find when h is zero. Table 2 could be rewritten as follows.

Table 3

t (time after midnight, hours)	0	2	4	6
h (height of bank above river, metres)	21	17	13	9

The graph of h against t would be as shown in Figure 18.

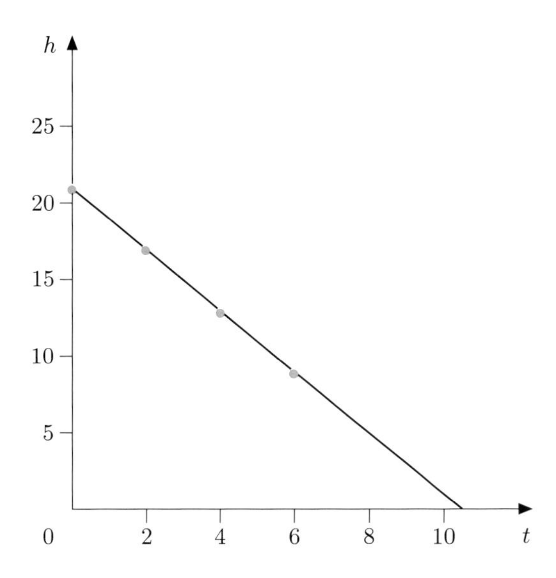

Figure 18

The intercept is 21. The gradient is $\dfrac{9-21}{6-0} = -2$. So the equation relating h and t is:

$$h = -2t + 21$$

The model is valid for $0 \le h \le 26$.

The problem is to find the time t when $h = 0$; that is, to solve the following equation:

$$0 = -2t + 21$$

So, rearranging, we have $2t = 21$, and hence $t = 10\frac{1}{2}$. So the river will reach the top of the bank at 10.30 am (as found before).

This illustrates an important point: there are sometimes several possible ways of defining the relevant variables. Assuming a constant rate of change (which may be positive or negative) leads to a linear graph (which may slope either way). A different choice of variables produces a different linear graph, although if the same assumptions have been made then the interpretation of the results should be the same.

Sometimes you may need to find the equation of a straight line without its intercept being obvious.

Example 5 Intercept unknown!

What is the equation of the straight line in Figure 19?

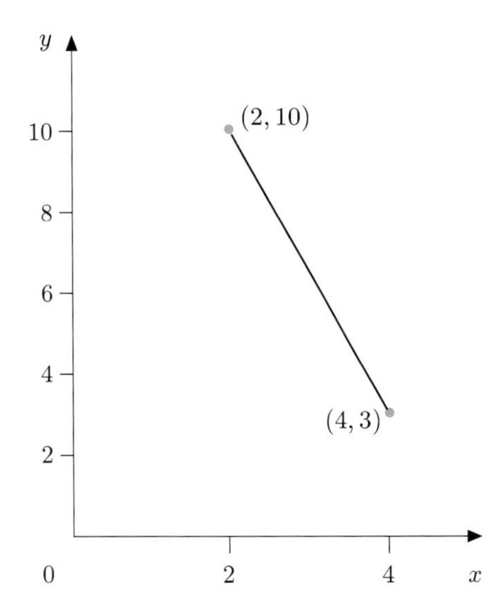

Figure 19

The line passes through the points $(2, 10)$ and $(4, 3)$. Hence the gradient of the line is given by:

$$\frac{\text{increase in } y}{\text{increase in } x} = \frac{3 - 10}{4 - 2} = -3.5$$

Hence the equation of this line is of the form $y = -3.5x + c$, where c is the intercept. There are several ways to find c. Here is one.

Because the line passes through the points $(2, 10)$ and $(4, 3)$, the equation must be satisfied at these points. Take either of them to find c (and then check you are correct with the other).

When $x = 2$, $y = 10$; so the equation $y = -3.5x + c$ becomes $10 = -7 + c$. Hence c must be 17 and the final equation is:

$$y = -3.5x + 17$$

(Check that the other point $(4, 3)$ satisfies this equation: $x = 4$ gives $y = -3.5 \times 4 + 17$, which is 3, as expected.)

There are several methods for finding the equation of a straight line. They all depend upon the following three main points.

Equation of a straight line

(a) The equation is satisfied for *all* points on the line, so the substitution of the coordinates of a known point into the equation should give equality.

(b) The gradient of a straight line is:

$$\frac{\text{increase in dependent variable}}{\text{increase in independent variable}}$$

It can be calculated between any two points on the line because the gradient of a straight line is constant. (Note that a decrease is regarded as a 'negative increase'.)

(c) The general equation of a straight line is of the form

$$y = mx + c$$

where y is the dependent variable, m is the (constant) gradient, x is the independent variable and c is the (constant) intercept (the value of y when x is zero).

Activity 12

What are the equations of the straight lines in Figure 20?

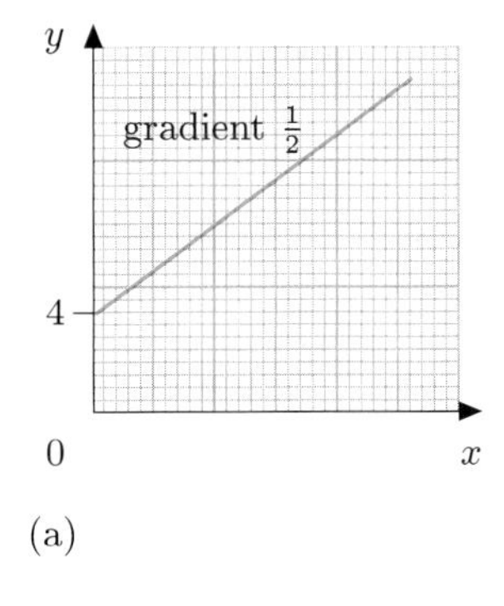

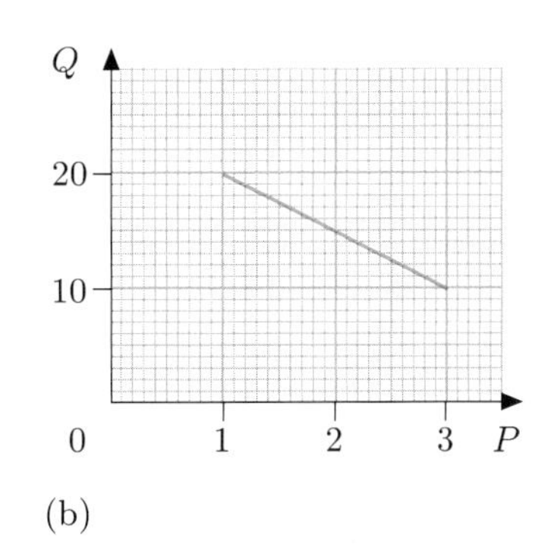

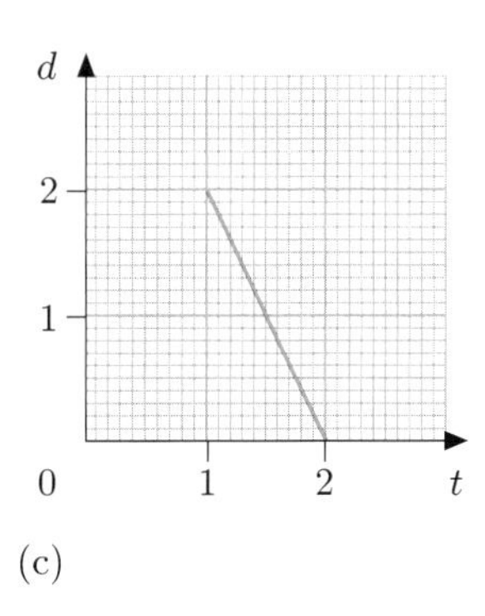

Figure 20

Activity 13 Families of straight lines

Use your calculator to plot the following sets of straight lines and describe each set in a few words, as if to a friend.

(a) $y = x$
$y = x + 1$
$y = x - 1$
$y = x + 2$

(b) $y = 2x$
$y = 2x + 1$
$y = 2x - 1$
$y = 2x + 2$

(c) $y = -x$
$y = -x + 1$
$y = -x - 1$
$y = -x + 2$

(d) $y = x + 2$
$y = 2x + 2$
$y = -x + 2$
$y = -2x + 2$

(e) The general equation of a straight line is $y = mx + c$. Use your answers to (a), (b), (c) and (d) to help you to describe in words the set of lines which all have the same value of m and the set of lines which all have the same value of c.

In the general equation $y = mx + c$ of a linear graph, the letters m and c are called *parameters*: for any particular equation they take specific numerical values, and so they are not variables like x and y.

2.4 Interpreting solutions

So far this section has concentrated more on creating a model than on interpreting the results. But the interpretation of results is important.

In the motorway journey of Example 2, a constant speed was assumed for the whole journey. This is obviously most unlikely, and so should be taken into account when telling the passenger when to expect the car, which is unlikely to arrive at exactly 8.26 am. So there is no point in giving the solution to such accuracy: to say 'between 8.20 and 8.30 am' is far more realistic. The accuracy to which it is meaningful to give a solution depends very much on how realistic the assumptions were initially.

Because you use a calculator which gives many figures when displaying a result, it is often tempting to quote all the figures. However, there is no point in giving more figures than are meaningful in the interpretation of your results. The same is true when reading from a graph. You should round your final results to a degree of accuracy which is meaningful in the light of the assumptions and the context. You may be used to rounding your answers to two decimal places (2 d.p.) or three significant figures (3 s.f.), but in modelling you should interpret the results to fit with the purpose. There is no point in giving a solution like 'There would have been 601.48 (2 d.p.) geese in the winter of 1953'. There is no such thing as 0.48 of a goose. 'About 600 geese' is a better answer, as only the first one or two figures are likely to be significant in predicting a bird population.

Activity 14

A student tackling Activity 9 on snow-ploughs drew an exceptionally large and accurate graph and wrote down in answer to part (d): 'Tell them to start snow-ploughing at $47\frac{1}{2}$ minutes past 3 in the afternoon'. Give a more sensible interpretation of this result.

Another point to bear in mind is that the accuracy of your results depends on the accuracy of your data. This is especially important when using a calculator. Always enter the data as accurately as you can and then *round off at the end.* The way in which you round up or down also may depend upon the purpose. 'If the next figure is five or more then round up and if it is a four or less round down' is the convention used to round to the nearest figure. However, there may be cases where you need to use different rounding techniques. For example, the people managing the flood warning and the snow-ploughs need to err on the side of caution and should round the predicted times down rather than up.

If you round intermediate results, the errors in doing this can accumulate.

Another useful device in interpreting solutions is to allow a safety margin. The authorities controlling the snow-ploughs would be wise to call the drivers out earlier, rather than later, in case the snow came down faster. The flood authorities would also be well advised to allow a reasonable safety margin. In practice of course other factors may be relevant, such as the cost of snow-plough drivers' pay and the desire not to evacuate people from their homes unnecessarily. A suitable balance has to be found.

As you saw in Activity 4, most functions approximate to a straight line over small ranges of values. Hence linear models are good short-range predictors of change. However, you need to be cautious in your interpretation of the results of using linear models as predictors over a long range.

Making predictions beyond the range for which data are available is called *extrapolation.* Both the snow-plough and flood-warning examples are instances of this. Making predictions within the range for which data are available is called *interpolation.* The motorway-journey example is an instance of this.

Activity 15 Demand for tomatoes

A greengrocer has kept a record of the quantity of tomatoes she sells as the selling price varies over several months, in order to try to predict the likely demand at different selling prices. She summarizes her sales in Table 4, and wants to use these results to help her decide how many tomatoes to buy from the market at different market prices.

Table 4

P (price per kg, pence)	20	25	30	35	40	50
Q (quantity sold per day, kg)	102	92	82	72	62	42

(a) Set up a graphical linear model to help her. (You should aim to draw a straight line through all the points given in the table.)

(b) Predict the weight of tomatoes sold per day if the price of tomatoes at the market means that the greengrocer will have to charge:

(i) 46 pence per kg

(ii) 15 pence per kg

(iii) 28 pence per kg

(iv) 68 pence per kg

(c) Interpret your solution to (b) as a recommendation to the greengrocer about the quantity of tomatoes she should buy on each occasion.

The greengrocer might be a little sceptical about the validity of any recommendation in the 68p case, because a price of 68p is rather higher than she has experienced before. In setting up the model, only data on prices of between 20p and 50p were available, and so she might understandably be cautious about extending the model to cover prices which are much beyond that range.

As you saw earlier, it is good idea to specify the range of values of each variable when setting up a model. So in the above model for tomatoes you might restrict P to lie between 20 and 50 ($20 \leq P \leq 50$) or perhaps between 10 and 60 ($10 \leq P \leq 60$). It is also a good idea to note any other limitations of the model. For example, the above model for tomatoes does not take into account factors such as seasonal fluctuation in demand.

Activity 16

Look back at the models listed below and indicate the limitations of each of them.

(a) Motorway journey (Example 2)

(b) Snow-ploughs (Activity 9)

(c) Flood warning (Example 3)

2.5 Using linear models

You now know how to create linear models both graphically and algebraically. This subsection provides your with more practice in using linear models. But first, here is a summary of the creation process.

Creating an algebraic linear model

The technique for creating an algebraic linear model can be summarized in five stages.

(a) Define the dependent and independent variables and represent them by symbols.

(b) Assume a constant rate of increase (gradient). Find a numerical value for it.

(c) Find the initial value (intercept) for the dependent variable (its value when the independent variable is zero).

(d) Write down the equation in the form:

$$\text{dependent variable} = \text{rate of increase} \times \text{independent variable} + \text{initial value}$$

(e) State the limitations of the model.

Although the above summary is for an *algebraic* linear model, do not forget that it is often useful to sketch the linear graph as well as use the linear equation. A linear graph is quick to draw and gives a visual understanding of how the variables are related, to complement the generally more useful linear equation.

Activity 17 Manufacturing specialized instruments

A firm produces a specialized instrument and has the facilities to produce 100 instruments per week. The firm's accountant estimates that the firm pays out £5000 per week on fixed costs (overheads, salaries, and so on) and that on top of this the cost per instrument is constant at £50 per instrument.

You are required to create a linear model to help the firm see how their total costs vary with the number of instruments produced.

(a) Sketch an appropriate graphical linear model and find its equation.

(b) State any limitations of the model.

(c) What do the intercept and the gradient of the graph represent?

(d) How much would it cost to produce 80 instruments in a week?

You have already seen how to use the values of one point to determine the unknown intercept in the equation of a straight line. The same technique can be applied to determine the initial value for a linear model.

Example 6 Demand for soft fruit

◁ *Specify purpose* ▷

As prices go up, demand for soft fruit usually falls off. One local producer has kept track of the quantity of fruit sold at the local shops supplied by him at different prices (Table 5) in order to predict the demand as the prices change.

Table 5

Price per kg (pence)	10	30	50
Demand per week (kg)	2000	1600	1200

The producer wants to predict the demand for soft fruit at different prices, in particular at 35, 40, 45 and 55 pence per kg. Set up an algebraic linear model to help him do this.

Solution

◁ *Create model* ▷

(a) Define the variables: let the quantity demanded be Q kg per week and let the price be P pence per kg.

(b) Assume Q decreases at a constant rate as P increases. Q decreases by 400 for an increase of 20 in P so the constant rate of increase is $-400/20$ (negative because Q is decreasing), which is -20.

(c) You do not have a starting value for Q. For the moment call it Q_0 (as it is the value of Q when P is 0).

(d) The equation is $Q = -20P + Q_0$, or equivalently:

$$Q = Q_0 - 20P$$

(e) The model will probably be valid in the range $5 \leq P \leq 55$.

◁ *Do the maths* ▷

You know that Q is 2000 when P is 10, and these values must satisfy the equation. Substituting them into the equation gives:

$$2000 = Q_0 - 200$$

So $Q_0 = 2200$. Hence the equation is:

$$Q = 2200 - 20P$$

Substituting various values of P into the equation gives the numbers in Table 6. (Note that the values of Q for $P = 10, 30, 50$ do agree with the original data.)

Table 6

P	10	30	35	40	45	50	55
Q	2000	1600	1500	1400	1300	1200	1100

◁ *Interpret results* ▷

Hence the model predicts a demand of 1500, 1400, 1300 and 1100 kg per week at prices of 35, 40, 45 and 55 pence per kg respectively. The results obtained from the model will be a rough guide, correct maybe to the nearest 100. The model ignores many other factors that might influence demand (for example, the weather) and so the producer should not expect the results to be very accurate. However, they may be good enough for planning purposes.

◁ *Evaluate model* ▷

Finding an unknown initial value for an algebraic linear model

(a) Use a symbol like Q_0 to represent the initial value in the equation.

(b) Substitute into the equation one known pair of values of the variables.

(c) Hence find the initial value.

(d) Check that the equation with this initial value is satisfied by any other data you have.

Activity 18 *Supply of soft fruit*

When the market price of soft fruit becomes too low, the producer feels that it would be uneconomical to employ anybody to pick the crop. However, at higher prices he can employ more pickers, and as the price rises it becomes economical to pick fruit in areas which were uneconomical at lower prices. At 30 pence per kg the producer reckons it would be economical to pick about 800 kg; at 50 pence per kg, 1600 kg could be picked economically.

(a) Set up an algebraic linear model for the quantity of fruit it is economical for the producer to supply at different prices.

(b) Use the model to predict how much fruit it is economical for the producer to supply when the market price is 35, 40 and 55 pence per kg.

As this section comes to an end, take some time to review what you have learned by completing Activity 19. Think also about how you 'read' the text.

Activity 19 *Defining terms*

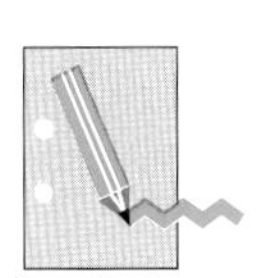

This section has introduced a number of terms associated with linear models. You should have defined them for your Handbook as you were working through the section. Review your learning for this section by looking back through your Handbook entries for this section.

Many of the terms you have included in your Handbook may be considered to be 'specialist' terms. In your view, when is it appropriate to use such terms in written and discussion work?

The main things to remember about mathematical modelling in the context of linear models can be summarized as follows.

Creating a linear model

(a) Determine the relevant quantities for independent and dependent variables and represent them by symbols. These should be the quantities which are directly relevant to the purpose. There may be more than one equally good way of doing this; the models produced will be different but the interpretation should be the same.

(b) Make the simplest realistic assumption; often this is to assume the rate of increase to be constant. Find a numerical value for this rate. It is the gradient of the linear graph.

(c) Look for an initial value (intercept)—if one is not obvious, represent it by a symbol.

(d) Set up the algebraic linear model in the form:

$$\text{dependent variable} = \text{rate of increase} \times \text{independent variable} + \text{initial value}$$

(e) If an initial value is needed, substitute a known pair of values for the variables into the equation in order to find it.

(f) Check that any other known pairs of values for the variables also satisfy the equation.

(g) Specify the limitations of the model. The data and information you have may be valid only for a certain period or under certain circumstances. It is a good idea to write down any limitations in order to remind you what they are when you come to interpret the solution.

Interpreting the results

(a) The mathematical results need to be interpreted in the light of the assumptions made and with reference to the purpose of the model. Note the limitations of the model, and treat with caution any results obtained by using the model beyond its limitations.

(b) Quote the results only to as many figures as are meaningful in the light of the assumptions and in the context of the real-life problem. Round the results up or down according to the context. Add in an appropriate safety margin if necessary, bearing in mind the purpose of the model.

Outcomes

After studying this section, you should be able to:

- ◇ create graphical and algebraic linear models from given data or written descriptions (Activities 8, 9, 15, 17, 18);
- ◇ find the gradient, intercept and equation of a given straight-line graph (Activities 10, 11, 12);
- ◇ interpret the gradient and intercept of a linear graph in terms of the graph and/or the modelling situation (Activities 11, 12, 13, 15);
- ◇ use a linear model to make relevant predictions interpreted in the light of the modelling assumptions and limitations and the purpose (Activities 8, 9, 14, 15, 16, 17, 18).

3 Linear models directly from data

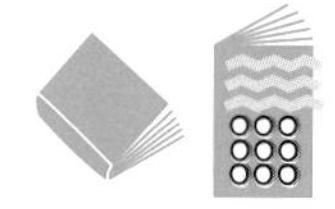

Aims This section aims to teach you to use the course calculator to find the 'best' linear model for a set of data and to plot the data points and the line on the same graph on your calculator. ◇

This section is concerned with finding a 'best fit' to a given set of data. Each data point will consist of a pair of numbers, and the 'best fit' will be a straight line. Such data pairs arise frequently as the outcome of surveys and experiments.

3.1 What is a 'best fit' line?

In modelling the demand for tomatoes in Activity 15, you discovered that one straight line went through all the points in the data set. Here are the data again.

Table 7

P (price per kg, pence)	20	25	30	35	40	50
Q (quantity sold per day, kg)	102	92	82	72	62	42

Here are two more data sets: they are for the motorway journey from Example 2 and for the demand for soft fruit from Example 5.

Table 8

Time taken (hours)	0	0.5	0.75	1	1.5
Distance travelled (miles)	0	29	43.5	58	87

Table 9

Price per kg (pence)	10	30	50
Demand per week (kg)	2000	1600	1200

In the case of Example 2, see Figure 9.

If, for each of the sets of points in Table 8 and Table 9, you were to plot the points on a graph, you would easily discern that, as for the points in Table 7, a straight line would go through all the points plotted.

Alternatively, scrutiny of the numbers in each table yields the same message: in Table 7, an increase of 5 pence per kg in the price resulted in a 10 kg reduction in the quantity sold per day; in Table 8, 29 miles are travelled with the passing of each half-hour; and in Table 9, a 20p increase in price leads to a 400 kg reduction in demand per week.

Section 2 discussed how to find the linear equation linking (that is, describing the relationship between) the dependent variable and the independent variable for such cases. It takes the form:

$$\text{dependent variable} = \text{rate of increase} \times \text{independent variable} + \text{initial value}$$

But what if the points in a data set do not lie exactly on a straight line? Is a linear model still valid? And if so, how can the appropriate linear equation be determined?

Example 7 Copper pipes expanding

In an experiment to discover how the type of copper pipes used in domestic hot water pipes expand when heated, the following data for the lengths of a piece of copper pipe at various temperatures were obtained.

Table 10

Temperature of pipe (°C)	20	30	40	50	60
Length of pipe (mm)	1000.8	1002.1	1004.1	1005.9	1007.3

The scatterplot in Figure 21 shows the five plotted points, where temperature (the variable controlled during the experiment) is plotted on the horizontal axis and pipe length on the vertical axis.

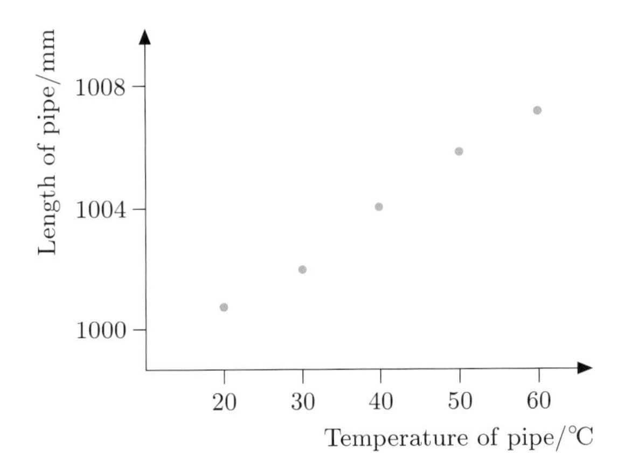

Figure 21

You can see that the points lie close to, but not actually on, a straight line. The data show 'perturbations'. These perturbations may be due to experimental error, or to measurement error; or maybe they exhibit a genuine physical phenomenon that expresses the relationship between pipe length and pipe temperature—evidently rather a complicated relationship!

Depending on the purpose to which such a model might be put, it might be useful to model the relationship between the two variables as a straight line—such a straightforward linear model might be rather useful, since the perturbations are very slight. (It is not claimed that the model would be 'correct'—just that it might be 'useful'.) The question then arises: how do you discover the 'best' straight line corresponding to the data points—the 'best' model?

Example 8 The tallest man

According to my 1984 copy of the *Guinness Book of Records*, 'the tallest man of whom there is irrefutable evidence was ... Robert Pershing Wardlow, born ... on 22 February 1918 in Alton, Illinois, USA'. The figures in Table 11 give his height, measured at various ages.

There are 12″ (12 inches) in 1′ (1 foot).

Table 11

Age (years)	5	9	11	13	15	17	19	21
Height (feet and inches)	5′4″	$6'2\frac{1}{2}''$	6′7″	$7'1\frac{3}{4}''$	7′8″	$8'0\frac{1}{2}''$	$8'5\frac{1}{2}''$	$8'8\frac{1}{4}''$
(centimetres)	163	189	201	218	234	245	258	265

Mr Wardlow died on 15 July 1940, aged 22. The scatterplot in Figure 22 shows the data points for height (on the vertical axis) against age (on the horizontal axis).

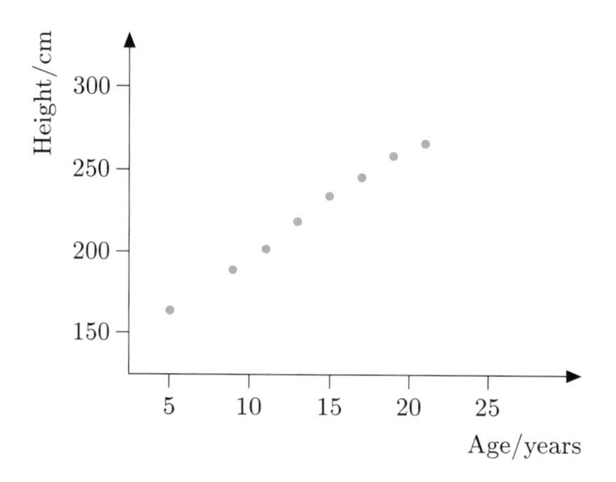

Figure 22

In this case a straight line fits the data points moderately well, though the fit is not as good as in the previous example, and it is certainly not perfect, as for the earlier data sets. This evidence of perturbation may be due to the fact that a natural process is involved (human beings do not grow according to a mathematical formula!), but notice also that no indication is given of when, during the year, this individual was measured.

The question again arises: if it is required to fit a straight line to these data points (assuming, that is, that some purpose to modelling growth with passing time by a linear equation has been identified) then what is the equation of the 'best' straight line corresponding to the data points?

The theory that gives the equation of the 'best fitting' straight line is moderately complicated, and so the details are not included here. (The course calculator knows all the necessary arithmetic, as it happens.) But the principle on which the theory is based is fairly easy to comprehend, and goes as follows.

Each of Figures 23 and 24 reproduces the length against temperature scatterplot derived from the data of Example 7, plus a suggested straight line corresponding to the data.

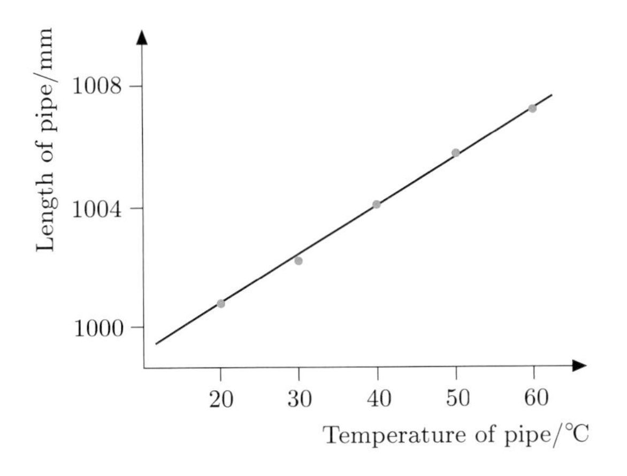

Figure 23

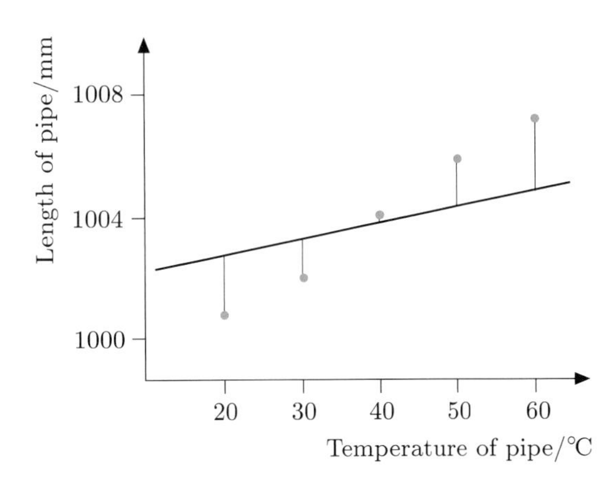

Figure 24

You can see that the straight line in Figure 23 fits the data rather well; in Figure 24 the fit is deliberately rather poor. The quality of the fit may be expressed in terms of the distances from the plotted data points to the fitted straight line, and these have been drawn in the figures. The smaller these distances, the better the fit. For reasons which are a little bit sophisticated (but they are good reasons!), the distance from each point to the line is measured in a vertical direction; and the overall quality of the fitted straight line is measured not by adding together the individual distances, but by adding together the *squares* of the distances. (Remarkably, this makes the mathematics involved easier.) The 'best' line is the line for which the sum of the squared distances is as small as possible, and it turns out that there is only one such line for any given data set. In fact, for the copper pipe data, it is the line shown in Figure 23.

The line is called by various names: the *least squares best fit line*, or simply the *least squares line*, or the *regression line*. Finding this line can be tedious if done by hand, but your calculator has the facility to calculate it for you.

Finally, but incidentally: if the data points are such that they really do lie on a straight line, then that line *is* the regression line for those points (all the point-to-line distances will be zero, as a rather special case).

Activity 20

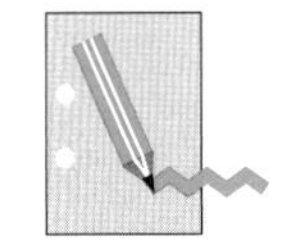

Describe, as if to a fellow student, the main principles behind the regression line as a 'best fit' straight line for a set of data.

3.2 Using the calculator for a 'best fit' line

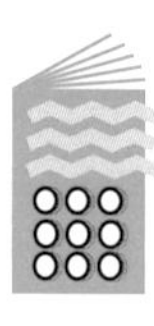

Now work through Section 10.1 of Chapter 10 of the Calculator Book.

Activity 21

Use your calculator and Tables 7, 8 and 9 to find the regression lines for the data for:

(a) the demand for tomatoes (Activity 15);

(b) the motorway journey (Example 2);

(c) the demand for soft fruit (Example 6).

In each case, explain the significance of the parameters given by your calculator for the regression line and check that all the data points lie on the line.

In cases (b) and (c), check that the equation of the regression line is the same as the linear equation derived at the beginning of Subsection 2.2 and in Example 6 respectively.

Activity 22 Copper pipes expanding

Input the copper pipe data from Table 10 into your calculator and plot the data points on a graph showing length L, measured on the vertical axis, against temperature T, measured on the horizontal axis. Use a vertical scale running from 990 mm to 1010 mm and a horizontal scale running from 0 °C to 80 °C.

Fit a regression line to this data. Plot the points and the line on your calculator screen. Explain the significance of the line's intercept and slope.

Activity 23 The tallest man

Assuming a straight-line model for the dependence of Robert Pershing Wardlow's height as a function of his age, calculate his average annual rate of growth as given by the regression line through the data points in Table 11.

The process of constructing a regression line relating two variables from given data which strongly indicate a linear relationship between them is quite straightforward. It is, however, worth noting that the evidence of half-a-dozen data points is scarcely strong enough to 'prove' that such a linear relationship actually exists. The truth of a law such as 'the expansion of copper pipe is proportional to the change in temperature' can only be confirmed by much more experimental evidence.

3.3 Interpolation and extrapolation

Once more consider the demand for tomatoes (Activity 15). As Activity 21(a) showed, the data in Table 4 can be fitted by the straight line

$$Q = -2P + 142$$

to predict values of Q (the quantity of tomatoes sold per day, in kg) for values of P (the price per kg, in pence) which did not appear in the table. For example, the demand at a price of 28 pence per kg is predicted to be:

$$Q = -2 \times 28 + 142 = 86$$

One may often infer values of a variable by fitting a model to the data and using it to calculate the required values. Recall that, when the required values lie within the range covered by the data, this is called *interpolation*.

More weight attaches to interpolation the more data points it is based on, and if there is reason to believe that the variables are related in a way which rules out large fluctuations unrevealed by the data. Annual seabird data from Skomer (*Unit 5*) are *not* suitable for interpolation, owing to large seasonal fluctuations, but the data from a copper pipe expanding or a person growing do appear to be suitable.

However, even when a linear relationship between two variables is well established for certain ranges of values of the variables (ranges covered by the available data), the relationship may not hold for values outside those ranges. Thus, although the data for the height of giant Robert Pershing Wardlow indicated a fairly convincing linear pattern of growth (Activity 23), it is unreasonable to expect that growth would have continued linearly had he lived beyond 22; if it had, his height at age 30 would have been nearly 11 feet! His height at age 0 (birth) is also unlikely to have been given by the intercept of the graph, at about 130 cm, that is 1.3 metres (over 4 feet).

Using 'best fit' lines for the prediction of values of a variable outside the range for which data are available is an example of *extrapolation*. The risks of error as a result of extrapolation are many: predictions of the future are often based on extrapolation and frequently fail. Statisticians have formulated a rough rule of thumb to use under these circumstances: never extrapolate further ahead or backward than one quarter of the period for which you have data, so that if you have annual data for 20 years, say, then extrapolate for no more than 5 years ahead or backward. This rule suggests an appropriate degree of caution to use when extrapolating from *any* purely empirical data.

An example in which extrapolated conclusions are unreliable but perhaps interesting is furnished by athletics. Way back in May 1954, many people were excited by the achievement of Roger Bannister in running the mile in less than 4 minutes for the first time. Nowadays, the mile is run in well under 4 minutes quite regularly. The 4-minute mile was a great psychological barrier in the early fifties; the next comparable barrier was the $3\frac{3}{4}$-minute mile.

Activity 24 Identifying key points

Read the reader article 'The ultimate mile' by Trevor Kitson.

A colleague has decided to make a video based on the article. She needs a summary of the key points described in the article so she can devise the different scenes. Summarize the key points for her.

The basic data show how the record improved between 1913 and 1984, when the article was written. A straight line was fitted to the data. The equation of the line is

$$y = -0.006\,933x + 4.358$$

where y is the current mile record in minutes and x is the date minus 1900. The line predicts that the $3\frac{3}{4}$-minute mile would be run in 1987 ($x \simeq 87.7$). Moreover, according to this extrapolation, the $3\frac{1}{2}$-minute mile will be run in 2023 ($x \simeq 123.8$).

The model is not very useful: the predictions are rather sensitive to the accuracy of the coefficients of the equation of the line, and past experience has shown that a record may stand unchanged for quite a long time. More seriously, if the line is extended all the way down until it crosses the x-axis, you find that in 2528 the mile will be run in no time at all, 'a feat which will presumably ruin athletics as a spectator sport', as Kitson points out. Clearly, this linear extrapolation is not very reliable. It is more likely that there is some limiting time for running the mile, below which it will be impossible to go, so that the graph must really be curved and not a straight line.

By fitting a curve, Kitson estimated this ultimate mile record. According to his calculation it would be 3 min 46.66 s, and would be run in 1998. The linear extrapolation predicts a mile record of 3 min 45 s in 1987. Both conclusions were equally well supported by the data. However, Steve Cram's 1985 record of 3 min 46.31 s supports the linear prediction rather than an 'ultimate mile' of 3 min 46.66 s.

▶ How does Nowedine Morcelli's 1993 record of 3 min 44.39 s fit in?

It still fits in reasonably well with the linear prediction. 1993 is 93 years after 1900 which from the linear equation gives a predicted record of:

$$y = -0.006\,933 \times 93 + 4.358 = 3.71 \text{ minutes (to two decimal places)}$$

The record of 3 min 44.39 s is 3.74 minutes (to two decimal places).

3.4 Accuracy of data

Whenever you use numbers derived from measurements, you have to settle for less then perfect accuracy. For example, it is difficult to measure your height more accurately than to within a few millimetres either way, or your weight to within half a kilogram or one pound. There are several reasons for this: for example, the ruler or scales used for measuring probably has no finer divisions, it is difficult to keep still while measuring, and the quantity is not very precisely defined (does 'height' include hair or shoes? does 'weight' include clothes?).

How accurately can you measure them?

Usually the number of significant figures given in data is an indication of the accuracy. This is a fairly standard convention. For example, if you read that the distance from Mercury to the Sun is 58 million kilometres, you are being told not that the distance is precisely 58 million kilometres but that it is closer to 58 million kilometres than to 57 or to 59 million kilometres. In other words, it lies between 57.5 and 58.5 million kilometres. Another way of saying the same thing would be to write that the distance from Mercury to the Sun is 58 ± 0.5 million kilometres. When the accuracy of a measurement is indicated in this way, in the form $x \pm \varepsilon$, ε is called an *error bound* for the measurement x. Thus the convention described above may be expressed by saying that *the error bound is roughly half a unit of the last digit shown.*

If the last digit shown is a zero, this convention is ambiguous. If somebody told you about a celestial body that was 90 million kilometres away, would the implied error bound be 0.5 million (treating the 0 as the last digit given, just like the 8 in the previous example) or 5 million (treating the 0 as just a place marker to distinguish 90 from 9)? Such ambiguities can be avoided by using *scientific notation*, where numbers are expressed in the following form:

$$a \times 10^b, \quad \text{where } 1 \leq a < 10 \text{ and } b \text{ is an integer (positive, negative or zero)}$$

In this notation there is no need for zeros as place markers and so every written zero is significant as far as accuracy is concerned. This enables '90 million' to be written in different forms to indicate different accuracies:

9×10^7 implies an error bound of 0.5×10^7
9.0×10^7 implies an error bound of 0.05×10^7
9.00×10^7 implies an error bound of 0.005×10^7

Unfortunately this convention is not always adhered to. So keep your wits about you, and always try to form your own estimate of accuracy.

Activity 25 *Accuracy of a census*

The population of London is given in a table as 6696 000. Express this in scientific notation under the assumption that:

(a) it is the exact result of a census;

(b) it is rounded to the nearest thousand.

Activity 26 *Your height and weight*

(a) Write down your height and weight, with estimates of the accuracy of your figures.

(b) Express your height and weight (as given in your answer to (a)) in scientific notation, in a form that indicates the accuracy of the values you gave. Now try to explain in words what this actually means.

(c) Give an estimate of the accuracy you might expect to achieve if you measured the length of time for your next journey as carefully as you could. Justify your answer.

Review your learning for this section by looking back through your Handbook entries for this section. Think also about how you 'read' the text.

This section explained the idea of a regression line as a 'best fit' line for a set of data and showed how the course calculator can be used to find it. It also discussed the dangers associated with using regression lines for prediction, particularly for extrapolation. The use of error bounds and scientific notation in describing the accuracy of data was also discussed.

Outcomes

After studying this section, you should be able to:

- ◇ describe the criteria for fitting a regression line to a set of data (Activity 20);
- ◇ obtain the equation of a regression line for a set of data pairs, using your calculator, and use this equation to make predictions (Activities 21, 22, 23);
- ◇ describe the accuracy of data (Activities 25, 26).

4 Simultaneous linear equations

Aims This section aims to teach you to use two simultaneous linear equations and to find their point of intersection, both graphically and algebraically. ◇

Sometimes it is helpful to use a model consisting of a pair of straight lines or linear equations. The point where these two lines intersect is often an important feature of the model. This section looks at some examples in the context of motion and economics. It looks at the graphical representations first, as these illustrate the principles involved, then goes on to apply these principles to algebraic representations.

4.1 Simultaneous graphical models

Example 9 Where is lunch?

A couple of years ago I went on holiday with a group of friends. We travelled in two vehicles: an old van, containing all our camping gear, and a newer faster car. The night before setting out, we discussed our plans. Those of us in the van planned to leave at 9.00 am and travel up the motorway at a steady 50 mph. Our friends in the car were planning to do some shopping and then leave an hour later, but to maintain a steady 70 mph. We wanted to meet for lunch and, so, our problem was where and when would the car catch up the van?

◁ *Create model* ▷ The variables are the distance y miles at a time x hours after 9.00 am.

The van's journey can be modelled by assuming that the van is travelling at a constant speed of 50 mph. The van starts at $x = 0$, $y = 0$; its motion is described by a straight line with gradient 50, passing through the origin $(0, 0)$, as shown in Figure 25.

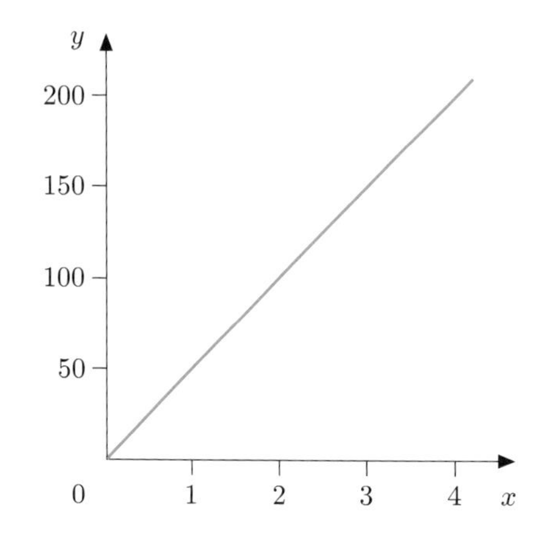

Figure 25

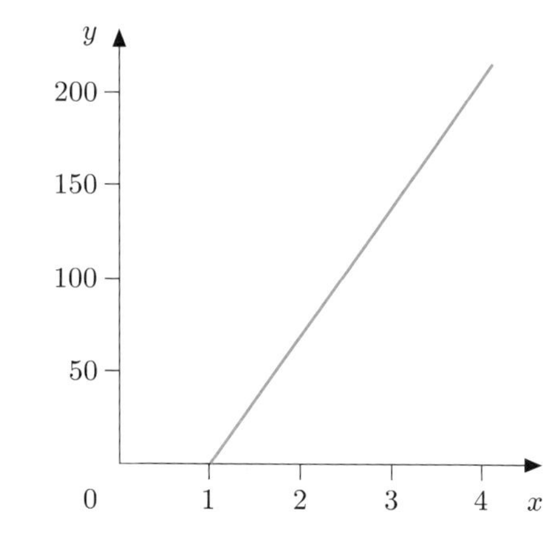

Figure 26

Assume that the car is also travelling at a constant speed (70 mph); but leaves one hour later. So its motion is described by a straight line with gradient 70, passing through the point $(1, 0)$, as shown in Figure 26.

To find where the lines intersect (that is, when both vehicles are at the same place at the same time), draw them on the same axes (Figure 27). The graphs cross at about the point $(3.5, 175)$: $x = 3.5$, $y = 175$.

◁ *Do the maths* ▷

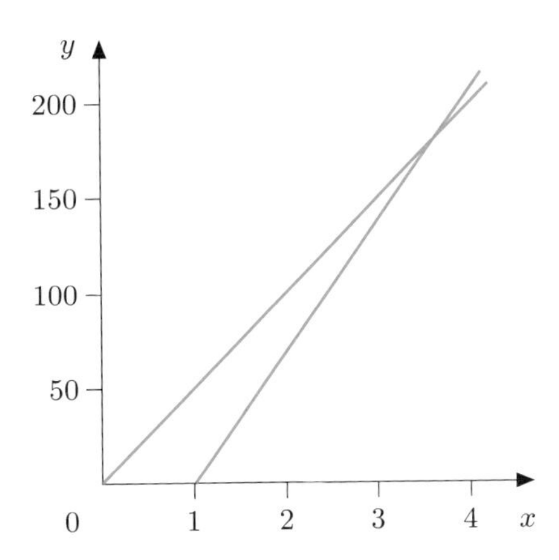

Figure 27

So the model predicts that we should plan lunch for about 12.30 pm ($3\frac{1}{2}$ hours after the van leaves), about 175 miles along the motorway.

◁ *Interpret results* ▷

The supply and demand of commodities may be estimated by surveys nationally (or even internationally) and economists use them to predict market prices. They find suitable models for the quantity that the producers are willing to supply at different prices and the quantity the consumers will demand to buy at different prices.

Example 10 Supply and demand of apples

Suppose that the demand for a particular type of apple is given in Table 12 and the supply of the same variety in Table 13. At what price will supply equal demand?

Table 12 Demand

P (price per kg, p)	Q (demand, 10^6 kg)
40	40
48	38
56	35
64	33
72	30
80	28
88	25

Table 13 Supply

P (price per kg, p)	Q (supply, 10^6 kg)
40	24
48	27
56	30
64	34
72	38
80	41
88	45

◁ *Create model* ▷ These points can be plotted on a graph and, if both demand and supply are assumed to change in a linear way with price, then two straight-line graphs result. If the data from Tables 12 and 13 are entered into your calculator, then the resulting regression lines are as shown in Figure 28.

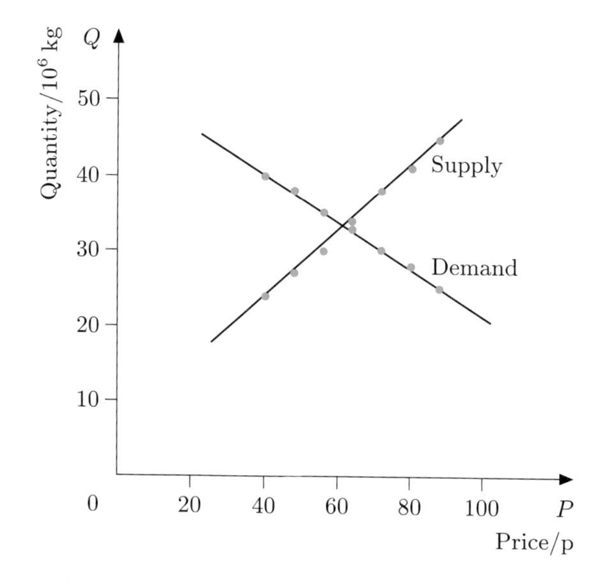

Figure 28

◁ *Do the maths* ▷ Using the trace facility of the calculator gives the point of intersection as $(62, 33)$, rounded to the nearest penny.

◁ *Interpret results* ▷ This point corresponds to a price of 62 pence per kilogram and to a quantity of 33 million kilograms.

The price at which supply equals demand, that is at which the suppliers are able to sell all they have produced (other things remaining equal), is the *equilibrium price* for the commodity. The term 'equilibrium' is used because, if the models for the dependence of supply and demand on price are good ones, a person selling at below this price would expect to have more customers than could be provided with apples, so the price might increase due to a shortage; while selling at above this price would lead to fewer customers than would exhaust the supply of apples actually available, so the price might fall due to a glut. Equilibrium would be achieved when supply equals demand.

Activity 27 Supply and demand of soft fruit

A survey suggests that the national demand for soft fruit varies with price according to Table 14. A second survey suggests that the supply of soft fruit available for sale varies with price according to Table 15.

Table 14 Demand

Price (pence per kg)	Quantity bought (10^6 kg)
15	30
20	26
25	23
30	19
35	16
40	12

Table 15 Supply

Price (pence per kg)	Quantity supplied (10^6 kg)
15	10
20	15
25	20
30	25
35	30
40	35

(a) Draw straight lines to show how the quantity of soft fruit demanded and the quantity supplied are expected to vary with price.

(b) Estimate the price at which the quantity of fruit demanded just equals that supplied. Estimate the quantity of fruit that is sold at that price.

You may use either graph paper or your calculator (using its regression, graphing and tracing facilities) for this activity.

This subsection has given you a visual idea of the principles involved in models comprising two linear graphs. If the straight lines each represent the motion of a vehicle (distance against time), the point of intersection predicts the place and time where they meet. In economic models of supply and demand, the point of intersection predicts the equilibrium price, where the quantity demanded equals the quantity supplied. In all cases it is important to choose the variables in the same way for both graphs, so that they can be superimposed and the point of intersection found.

4.2 Simultaneous algebraic equations

Graphical methods can be perfectly adequate for solving problems such as those in the previous subsection, but are often more time-consuming and less accurate than corresponding algebraic methods. This subsection considers how the examples of the previous subsection could have been tackled using models consisting of two linear equations, called *simultaneous equations*, and in particular how to solve such pairs of equations to find values of the variables equivalent to the point of intersection of the graphs. You may find it helpful to keep in mind the corresponding graphical representations as you go along.

Consider again Example 9. As before, let the distance be y miles at a time x hours after 9.00 am (when the van sets out). Assume as before that both vehicles travel at constant speed, the van at 50 mph and the car at 70 mph. They will be at the same place simultaneously at the point of intersection of the two lines represented by the corresponding equations of motion of the two vehicles.

The van travels at a constant speed of 50 mph and starts at $y = 0$, $x = 0$, so the linear equation giving distance travelled as a function of time is:

$$y = 50x \qquad (x \geq 0) \tag{2}$$

The car travels at a constant speed of 70 mph. There is no initial value given for y, so for the moment call it y_0. The equation representing the car's motion is:

$$y = 70x + y_0$$

The car sets out at $y = 0$, $x = 1$. Substituting these known values into the equation gives:

$$0 = 70 \times 1 + y_0$$

So $y_0 = -70$ and the equation representing the motion of the car (distance travelled as a function of time) is

$$y = 70x - 70 \qquad (x \geq 1). \tag{3}$$

▶ Plot equations (2) and (3) on your calculator, and use the trace facility to find where they intersect.

You should have found that the point of intersection is at about $(3.5, 175)$, as before.

Graphical methods are often perfectly adequate for solving pairs of simultaneous equations, but algebraic methods are usually more appropriate.

Equation (2) represents the van's journey and equation (3) represents the car's journey. At the time and place where the two vehicles meet, both equations are satisfied simultaneously. The value for y tells us how far away the meeting place is. The value for x tells us when the two vehicles reach it. Mathematically, the problem is usually expressed as:

Solve the simultaneous equations

$$y = 50x \tag{2}$$

$$y = 70x - 70 \tag{3}$$

There are two equations and two unknown quantities y and x. In order to find either y or x, you need *one* equation involving only *one* of the unknown quantities. The two equations need to be manipulated to obtain a third equation involving only one unknown quantity—that is, one of the unknowns must be *eliminated*.

In this case it is reasonably easy to eliminate y. Since y is $50x$ (equation (2)) and y is also $70x - 70$ (equation (3)) then $50x$ must be equal to $70x - 70$. This is called '*substituting* for y', and gives:

$$50x = 70x - 70 \tag{4}$$

Rearranging equation (4) gives:

$$70 = 20x$$

Therefore:

$$x = 3\tfrac{1}{2}$$

The corresponding value of y can now be found from either equation (2) or equation (3). Choosing equation (2) and substituting $x = 3\frac{1}{2}$ into it gives:

$$y = 50x = 50 \times 3\tfrac{1}{2} = 175$$

Hence the solution is $x = 3\frac{1}{2}$, $y = 175$.

This is the same solution as was obtained graphically.

Check that these values also satisfy equation (3): using $x = 3\frac{1}{2}$ gives $y = 70 \times 3\frac{1}{2} - 70 = 175$, which is correct.

There are other possible techniques for the solution of simultaneous equations; if you have met another method which you prefer, then do use that method. The rest of the subsection provides more examples of the solution of simultaneous equations. (Do not worry about what the letters stand for.)

Example 11 Solving simultaneous equations

Solve the following pair of simultaneous equations: that is, find the values of A and B which simultaneously satisfy them.

$$7 + B = 2A \qquad (5)$$

$$5 - B = A \qquad (6)$$

Rearrange one equation to get A in terms of B. Equation (6) is easier.

From equation (6), $A = 5 - B$, so substitute $5 - B$ for A in equation (5), giving:

$$\begin{aligned} 7 + B &= 2(5 - B) \\ 7 + B &= 10 - 2B \\ B &= 3 - 2B \\ 3B &= 3 \end{aligned}$$

So $B = 1$.

Substituting this into equation (6) gives $A = 4$.

Check that the values $A = 4$, $B = 1$ satisfy equation (5): $7 + 1 = 2 \times 4 = 8$.

Note there are other ways of getting to the solution. You could, for example, rearrange equation (5), to get

$$A = \frac{7}{2} + \frac{B}{2}$$

and put this into equation (6), giving

$$5 - B = \frac{7}{2} + \frac{B}{2}$$

which reduces to $B = 1$ again. Or you could rearrange equation (5) to get $B = 2A - 7$, and substitute for B in equation (6).

Activity 28

Solve the following pair of simultaneous equations.

$$X = 7 + Y \quad (7)$$
$$2X = 2 + Y \quad (8)$$

Occasionally a pair of simultaneous equations has no solution. Look at the following two equations:

$$y = 3x + 2 \quad (9)$$
$$3y = 9x + 5 \quad (10)$$

There are no values of y and x which can satisfy equation (9) *and* equation (10) simultaneously. The equations in fact correspond to a pair of parallel lines, which explains the lack of a solution—parallel lines do not intersect.

▶ Plot these two lines on your calculator and check this.

The two lines are shown in Figure 29.

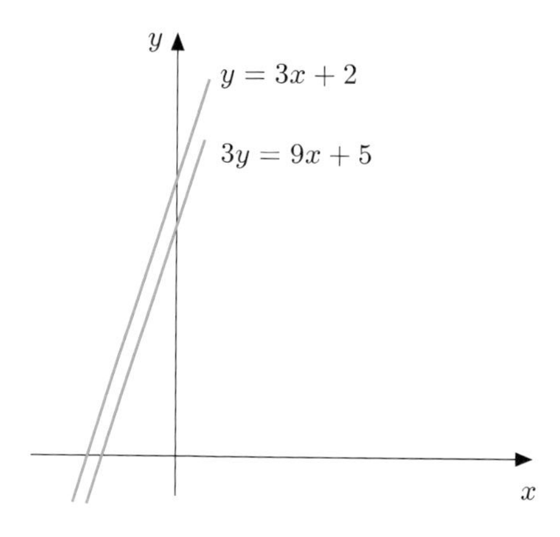

Figure 29

Sometimes, on the other hand, a pair of simultaneous equations has many solutions. Look at the following two equations:

$$y = 3x + 2 \quad (11)$$
$$3y = 9x + 6 \quad (12)$$

$y = 5$, $x = 1$ is one possible solution; $y = 2$, $x = 0$ is another; $y = -1$, $x = -1$ is another; and so on. The reason is that equation (12) is just equation (11) multiplied by three. Both equations correspond to the same straight line, and so any point on the line will satisfy both equations.

▶ Plot these functions on your calculator and check that you get just one line.

The resulting line is shown in Figure 30.

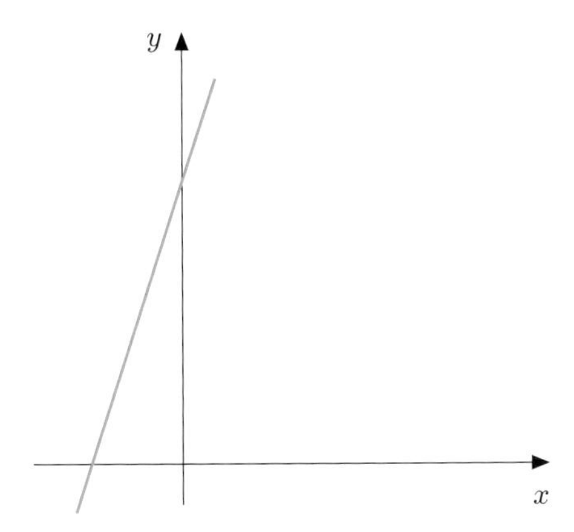

Figure 30

Notice that in both cases the trouble occurs when the coefficients of y and x in the second equation are the same constant multiple of those in the first equation. If you get such equations in a model, then it could mean that you have set up the model incorrectly. So you should go back and check your work.

Here is a summary of the substitution technique for solving pairs of simultaneous equations.

Solving simultaneous equations by substitution

(a) Rearrange one of the equations, if necessary, so that one variable is given in terms of the other.

(b) Substitute for this variable in the other equation.

(c) You now have one equation involving one unknown variable. Solve it to find the value of this unknown variable.

(d) Substitute this value into one of the original equations to find the value of other unknown variable.

(e) Check that this solution satisfies the other original equation.

Activity 29

Solve the following pairs of simultaneous equations.

(a)
$$2X = Y - 1 \quad (13)$$
$$3X = Y + 1 \quad (14)$$

(b)
$$X = 2Y - 3 \quad (15)$$
$$3X = -2Y + 15 \quad (16)$$

(c)
$$2A = 5B + 2 \quad (17)$$
$$2A = 3B - 2 \quad (18)$$

(d)
$$3P = 3 + 3Q \quad (19)$$
$$7P = 1 + 4Q \quad (20)$$

4.3 Simultaneous algebraic models

This subsection pulls together many of the ideas that you have met so far in this unit by going through the whole modelling cycle for a model involving a pair of linear equations: first create a model involving two linear equations; then solve the simultaneous equations; and finally interpret the solution.

Example 12 Best buy in belts

◁ *Specify purpose* ▷

A factory has a large conveyor-belt system and needs to replace worn belts regularly. A new type of belt has just been produced and the factory is testing a sample in order to decide whether or not they are a better buy than the older ones. The old belts each cost £1000 and involve maintenance costs of about £40 per month. On average they have a useful life of 20 months. The new belts are more expensive at £1400 each, but maintenance costs are half those of the old belts. The tests are not complete on the new belts, but they have been running for 20 months and still appear to have plenty of useful life in them.

Can any conclusion be made about which is the better buy?

◁ *Create model* ▷

Start by defining the variables: let the total cost of one belt be £C after t months of use. Assume that maintenance costs are constant each month on both types of belt. Ignore other costs like interest on capital and installation costs.

The model is only valid for the known useful life of the belts, that is for $0 \leq t \leq 20$.

To find which type of belt gives the lowest total cost overall, consider the value of C for each type of belt for different values of t. In particular, a break-even point is when C and t are the same for both types of belt.

Old belts

The purchase price of £1000 is the initial value for £C and the maintenance costs of £40 per month represent a constant rate of spending. Therefore:

$$C = 40t + 1000 \qquad (21)$$

New belts

Again the cost is made up of the purchase price (£1400) plus maintenance costs (£20 per month). Therefore:

$$C = 20t + 1400 \qquad (22)$$

So we have a pair of simultaneous equations:

$$C = 40t + 1000 \qquad (21)$$
$$C = 20t + 1400 \qquad (22)$$

◁ *Do the maths* ▷

Substituting for C from equation (21) into equation (22) gives:

$$40t + 1000 = 20t + 1400$$
$$20t = 400$$
$$t = 20$$

It is useful to check that both equations give the same break-even cost when $t = 20$, in case of error:

equation (21) $\quad C = 40t + 1000 = 1800$

equation (22) $\quad C = 20t + 1400 = 1800$

So the costs break even when $t = 20$.

◁ *Interpret results* ▷

The model predicts a break-even time of 20 months of use. Because the old belts now need replacing but the new ones appear to have more useful life in them, the new ones would appear to be a better buy.

◁ *Evaluate model* ▷

Before coming to a final conclusion, you should look at how realistic the model is. Among the factors to check are: whether interest charges on the extra £400 initial purchase price are negligible; whether any other factors (such as installation costs) should have been taken into consideration; and whether assuming constant maintenance costs was realistic. Also check such things as whether both types of belt are equally satisfactory and whether there are any price increases in the pipeline.

Activity 30 Aid for microprocessor development

A company is investing in a design project to build a microprocessor-based device. An aid to developing programs for the microprocessor is needed as part of the project, but there are several possibilities for obtaining such aid.

1. The company can buy a sophisticated program development aid for £8000 and employ two programmers to work with it.
2. The company can buy a less sophisticated program development aid for £1500 and employ three programmers to work with it.
3. The company can rent the more sophisticated aid at £850 per month and still employ the two programmers (the less sophisticated aid is not available for hire).

The running costs of the cheaper aid are estimated at about £40 per month, whereas those of the more sophisticated one are about £30 per month. Maintenance costs on the cheaper aid are likely to be £100 per month, whereas those on the sophisticated aid are likely to be £120 per month. The cost of employing a programmer is currently £800 per month (including overheads). The company is not sure exactly how long the project will last, but estimates suggest ten to twelve months.

(a) Construct models for the total cost (including purchase price) of this part of the project under each of the possible options.

(b) Compare options 1 and 3. Over what period of time would option 3 be more economical than option 1?

(c) Similarly compare options 1 and 2, and options 2 and 3. Over what period of time would option 2 be more economical than option 1, and over what period would option 3 be more economical than option 2?

(d) Consider the assumptions made in your models and interpret your solutions accordingly. What advice would you give the company?

Just as with ordinary linear models, involving just one straight line, care needs to be taken not to extrapolate too far when using simultaneous linear models. Making predictions too far outside the range of available data is unreliable.

Activity 31 Will women outrun men?

Read the reader article 'Will women soon outrun men?' by Brian J. Whipp and Susan A. Ward.

Explain the basis of their models, as if to a fellow MU120 student, and evaluate their models.

It is interesting to note that the linear model in 'The ultimate mile' article was based upon the time to run the mile, whereas here it is based upon the velocity (or speed), which is modelled as linear. In both cases, though, extrapolation of the linear model is not realistic indefinitely. A curve of some sort would be a better long-term model.

Using parameters

In Example 12 and Activity 30, a change in one of the prices quoted would have entailed repeating the calculations. This occurs quite often in modelling, and the problem of having to repeat much of the mathematics in such circumstances is overcome by using letters instead of numbers. These letters represent constants rather than variables and are known as *parameters*. A general model which could apply in several situations may have several parameters. The model is calculated only once, in general, using the parameters instead of numbers, and it can be evaluated for use in any particular situation by substituting appropriate values for the parameters.

You met parameters before in Subsection 2.3, in the context of the equation of a straight line.

Models that contain parameters cannot be solved by plotting graphs on your calculator. It is necessary to use algebra.

Example 13 Rent or buy?

A piece of equipment costs C to buy and running costs are at a rate R per time unit. However, a similar piece of equipment can be hired at a charge of H per time unit, which includes running costs. When is it more economical to buy and when to rent?

◁ *Specify purpose* ▷

Set up a model by first defining the variables. After time t, suppose the cost is K. The assumptions implicit in the problem are that the parameters C, R, H are positive constants, with $H > R$, and that there are no other costs; t is limited to the length of project or useful life of the equipment.

◁ *Create model* ▷

The problem is one of comparing the cost K at different times t for hiring and buying. In particular, is there a break-even point when the value of K is the same in both cases for the same value of t?

For 'buying' the costs are given by:

$$K = Rt + C \qquad (23)$$

and for 'hiring' (assuming no initial outlay at all) by:

$$K = Ht \qquad (24)$$

To find when the two costs are equal, you need to solve the simultaneous equations for K:

◁ *Do the maths* ▷

$$K = Rt + C \qquad (23)$$

$$K = Ht \qquad (24)$$

Substituting from equation (24) into equation (23) gives:

$$Ht = Rt + C$$

Subtracting Rt from both sides gives:

$$Ht - Rt = C$$

which can be written as:

$$(H - R)t = C$$

Check by multiplying out: $(H - R)t = Ht - Rt$.

Rearranging gives:

$$t = \frac{C}{H - R}$$

Find K by substituting into equation (24):

$$K = Ht = \frac{HC}{H - R}.$$

So the break-even point is after a time of $C/(H - R)$ at a price of $HC/(H - R)$.

Both equations (23) and (24) have straight line graphs. Their y-intercepts are C for equation (23) and 0 for equation (24). So, initially, buying (equation (23)) is more expensive than hiring (equation (24)).

◁ *Interpret results* ▷ If the equipment is required for some time less than $C/(H-R)$ then it is cheaper to rent; otherwise it is cheaper to buy.

◁ *Evaluate model* ▷ However, things like resale value may need to be included (this effectively reduces C) before a final decision can be taken in a particular situation.

When you are dealing with all-letter equations, it is easy to make a slip in the algebra. One way of checking this is to look at the types of quantities in the equation and in the solution and to see if they make sense. For instance, if one side of the equation is a quantity of money and the other is a quantity of money squared, you know something is wrong. Look at equation (24) in Example 13:

$$K = Ht \qquad (24)$$

What sort of quantities are K, H and t? K, the total cost, is a sum of money; t is a time; and H is a hire charge, which would be expressed as a sum of money per renting period. So H is money divided by time; and so Ht is a sum of money. Hence K and Ht are both sums of money.

Review your learning for this section by looking back through your Handbook entries for this section. Think also about how you 'read' the text.

To summarize, this section has covered situations which can be modelled by two straight-line graphs or two linear equations. In the example of the motion of two vehicles, the point of intersection of the lines and the solution of the simultaneous equations represents the vehicles' meeting place and time. In demand and supply examples, the intersection of the lines and the solution of the simultaneous equations represents the equilibrium price where the suppliers supply the same quantity as consumers demand. When setting up models of this kind it is important to choose the variables in the same way for both graphs and equations.

Such models can either be solved graphically (using your calculator) or algebraically. The algebraic technique of substitution for solving simultaneous equations involves manipulating the equations to obtain one equation in one unknown. This can be solved to find the value of one variable, which is substituted into one of the original equations to find the value of the other. This technique also applies to all-letter equations, where parameters represent constants which may change from one situation to another, for which graphical techniques are inappropriate.

Outcomes

After studying this section, you should be able to:

- ◇ solve simultaneous linear equations graphically and algebraically (Activities 27, 28, 29);
- ◇ model situations by simultaneous equations and comment on other people's models of this type (Activities 30, 31).

5 The most profitable strategy

Aims This section aims to widen your library of mathematical models to ones involving inequalities as well as equalities, and to teach you a strategy for predicting optimal strategies for some commercial situations. ◇

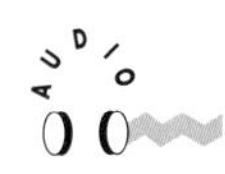

Although this section introduces some important mathematical ideas, it is one that can be skimmed if you are running short of time.

5.1 Equality and inequality models

All the models you have been using so far have been based on equations. Equality may not always be appropriate though. You may estimate that your speed 'will not be more than' 70 mph or the rate at which the river rises may be 'at least' 2 metres per hour. These translate into mathematics as inequalities.

In commercial operations inequalities are often appropriate models. For example, a firm may put a ceiling on spending on stock (total cost of stock must be less than or equal to a specified sum) or space may be limited (there is room for at most a specific volume of goods).

First consider a commercial problem with equalities before moving on to modelling a similar situation with inequalities.

Activity 32 Ice-cream sales

If you visit a local petrol filling station nowadays you are likely to find that it is a small convenience store as well, with displays of snacks, dairy goods, soft drinks and other frequently bought items. There is usually a selection of frozen foods as well. But, because of limited space, the manager may have a problem deciding which frozen foods to stock. Usually the problem involves many products. However, consider the simplified case of just two ice-cream products. Suppose the manager always fills one shelf of the frozen-food display cabinet with ice cream. Each week there is a delivery of two brands, Kreemy and Yummy, and past experience tells the manager that overall sales of 30 cartons can be expected each week. However, Kreemy is a 'premium' (expensive) brand and uses fancy packaging, so that each carton takes up twice as much room as the Yummy brand. In fact, a carton of Kreemy occupies one litre of space, and all together there are only 20 litres of storage space available. The manager's problem is to work out how many cartons of each type should be stocked so as to fill the available storage space and to meet demand. Experience has shown that there is no problem if one brand runs out—there are enough undiscerning customers to ensure that the 30 cartons are sold by the end of the week.

Can you help him determine what stock levels to adopt?

Problems such as the one in the previous activity can be extended to any number of variables. However, it is advisable to have a computer to solve more than two simultaneous equations!

◁ *Specify purpose* ▷

In Activity 32, the manager of the convenience store wished to balance conflicting requirements—total sales of the two brands of ice cream were 30 per week, but total storage space was limited. Suppose the manager's assumption that sales would be exactly 30 cartons per week was rather optimistic. In practice, the best that could be predicted, on the basis of past sales, was that *at most* 30 cartons would be sold.

◁ *Create model* ▷

If x is the number of Kreemy cartons and y the number of Yummy cartons sold, then instead of equation (32) in the comment on Activity 32 you have the *inequality*

$$x + y \leq 30 \qquad (25)$$

which states that total sales do not exceed 30 cartons.

Note that every inequality can be expressed in two ways: the statement $x \geq 0$ is the same as $0 \leq x$.

Notice that it does not make sense to talk about negative numbers of cartons sold, so both x and y are greater than or equal to zero: $x \geq 0$, $y \geq 0$. Now, recall from *Unit 7* that, when drawing graphs, the x-axis and y-axis create four regions, known as quadrants. The pair of inequalities $x \geq 0$, $y \geq 0$ defines the first of these quadrants, often referred to as the *upper right quadrant*, as shown in Figure 31.

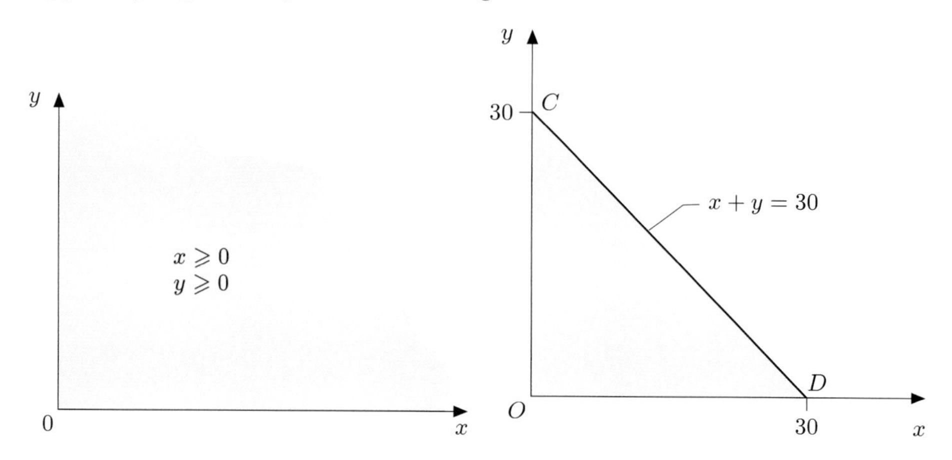

Figure 31 The upper right quadrant *Figure 32*

◁ *Do the maths* ▷

Now think about the graphical interpretation of inequality (25). When the 'equals part' holds then you have the straight line $x + y = 30$, shown as CD in Figure 32.

▶ Think of some values of x and y for which $x + y < 30$.

There are many values of x and y satisfying the condition $x + y < 30$. Some are shown in Table 16.

Table 16

x	y	$x+y$
0	0	0
1	0	1
20	2	22
3	3	6

▶ Plot these points and other ones you thought of on Figure 32.

You should have found that they all lie inside the shaded region in the figure.

However, if you try $x = 30$, $y = 30$, then this point has $x + y > 30$ and lies outside the shaded region.

Points on opposite sides of the line CD satisfy opposite inequalities. The shaded region OCD in Figure 32, including its boundary lines, represents the inequality $x + y \leq 30$. The remainder of the upper right quadrant represents the inequality $x + y > 30$. Figure 33 illustrates this.

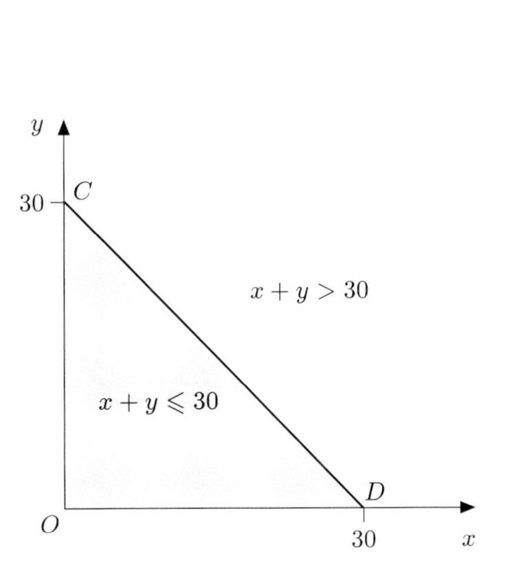

Figure 33

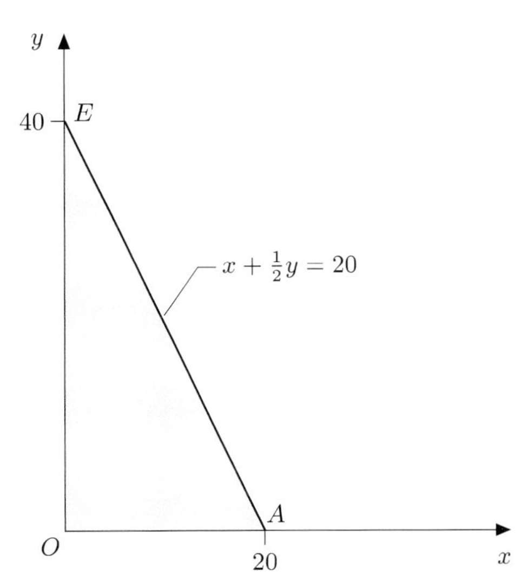

Figure 34

◁ *Specify purpose* ▷

Previously, the manager's objective was to fill up one shelf of the frozen-food display cabinet with ice-cream cartons. Perhaps, a more practical objective would be not to overfill the shelf.

◁ *Create model* ▷

So, remembering that the Kreemy cartons occupy 1 litre of space and the Yummy cartons $\frac{1}{2}$ litre, and that the storage space is *at most* 20 litres, equation (33) in the comment on Activity 32 needs to be replaced by the following inequality:

$$x + \tfrac{1}{2}y \leq 20 \qquad (26)$$

◁ *Do the maths* ▷

Figure 34 shows the line $x + \frac{1}{2}y = 20$.

▶ Which side of the line represents $x + \frac{1}{2}y \leq 20$? The shaded or the unshaded region? Try out some points if you are unsure.

This inequality corresponds to the shaded region in Figure 34, including the boundary lines OA, AE, EO.

◁ *Create model* ▷

Now put together the two shaded regions in Figures 32 and 34, as in Figure 35. The shaded area that they have *in common* is the region $OABC$, which represents the area where x and y satisfy both inequalities: both the restriction on total sales $x + y \leq 30$ and the restriction on storage space $x + \frac{1}{2}y \leq 20$.

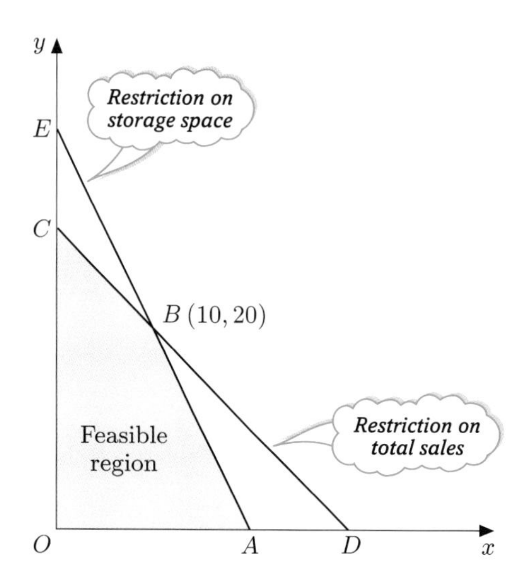

Figure 35

So this region contains all the points which simultaneously satisfy *both* of the inequalities (25) and (26). For this reason the shaded region (including its boundary lines) is called the *feasible region*. The inequalities are referred to as the *constraints*, because they 'constrain', or restrict, the opportunities for action.

◁ *Specify purpose* ▷

There is no longer a unique solution to the problem as it presently stands: all the points in the feasible region represent solutions. However, one thing has been overlooked that is undoubtedly there in the real world—the manager will wish to maximize the profit on ice-cream sales!

◁ *Create model* ▷

Suppose Kreemy—the premium brand—carries 30p profit per carton, whereas Yummy only brings 10p profit per carton. The overall profit P (in pence) when x Kreemy cartons and y Yummy cartons are sold is:

$$P = 30x + 10y \qquad (27)$$

The manager's problem is therefore to find the values of x and y, lying within the feasible region in Figure 36, which make the value of the profit P in equation (27) as large as possible.

◁ *Do the maths* ▷

Notice that if P in equation (27) is an actual number, say 200, then you get

$$200 = 30x + 10y$$

which is the equation of another straight line, shown in Figure 36.

Since the boundaries of the feasible region, shown in Figure 35, are straight *lines*, and the profit also corresponds to a straight *line*, the problem faced by the manager is an example of what is called a *linear* programming problem.

> ### Why linear 'programming'?
>
> The term *linear programming* was invented in 1951 when the term 'programming' was very new and trendy. Computer programming was still a novelty, and terms like 'linear programming', 'mathematical programming', 'quadratic programming', 'dynamic programming' and 'integer programming' were all coined around the same time to emphasize their newness and 'high-tech' connotations. To solve real-life problems in these areas, computers are usually essential, and even the relatively primitive machines of the 1950s were able to solve quite large linear programming problems. The other 'programming' problems had to wait for the much more powerful computers of later decades before large problems could be solved.

In Figure 36, the line with $P = 200$ cuts the x-axis at $x = 6\frac{2}{3}$ and the y-axis at $y = 20$.

Suppose P in equation (27) increases to 300—you get the straight line

$$300 = 30x + 10y$$

which is shown in Figure 37.

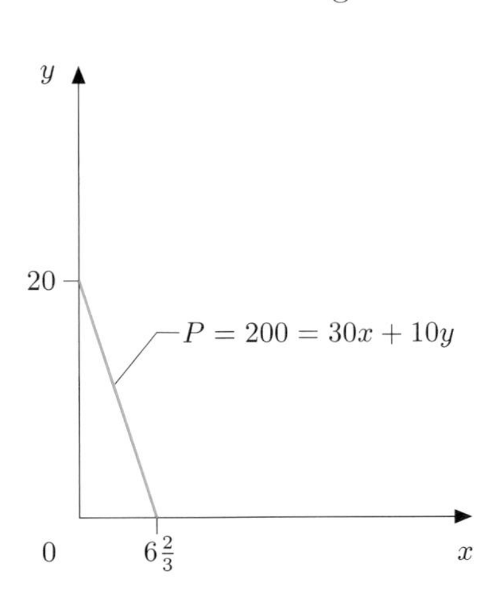

Figure 36

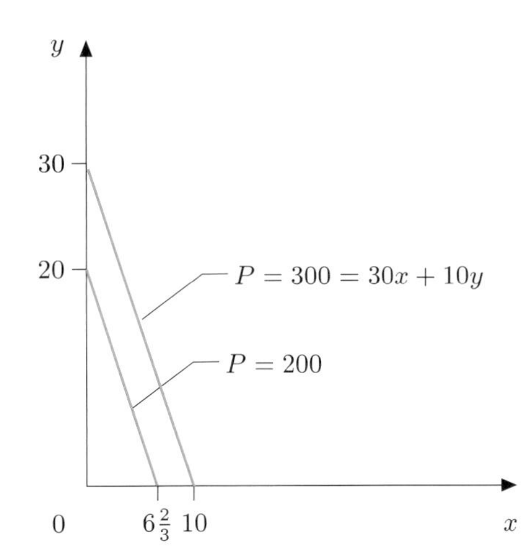

Figure 37

This line cuts the x-axis at $x = 10$ and the y-axis at $y = 30$. The two 'P-lines' in Figure 37 are *parallel* to each other but the second is *further away* from the origin.

The profit P in equation (27) is represented by a series of parallel straight lines which get further away from the origin as P increases in size. The manager's problem, in graphical terms, is to find a point, within the feasible region, on a 'P-line' which is the furthest from the origin—that is, produces the largest profit.

The problem can be solved as shown in Figures 38 and 39. In Figure 38 the two lines $P = 200$ and $P = 300$ are superimposed on the feasible region from Figure 36.

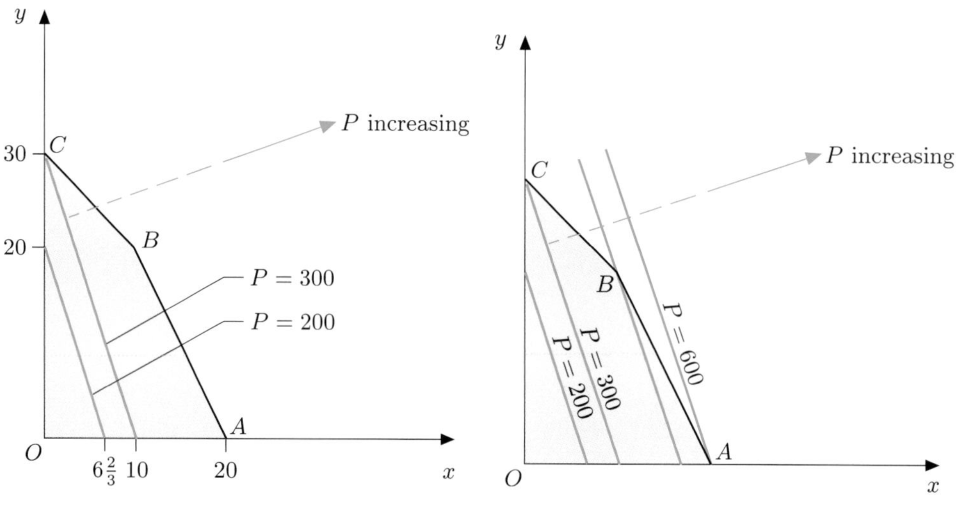

Figure 38 *Figure 39*

Imagine sliding one of these lines away from the origin, parallel to itself, in the direction of increasing P. You will reach the point B, but keep going—the line remains partly inside the feasible region until you reach the point A (as Figure 39 illustrates). Here you must stop, as going any further will take the P-line completely outside the feasible region.

The point A is inside the feasible region and has coordinates $x = 20$, $y = 0$. This is the point in the feasible region which is furthest away from the origin in the direction of increasing P.

◁ *Interpret results* ▷

In other words, the manager's best option is to stock 20 cartons of Kreemy and none of Yummy. The total profit will then be $30 \times 20 = 600$p. Notice that this is achieved with a total sale of 20 cartons, less than the maximum possible demand, but on the other hand the full storage capacity is utilized.

◁ *Specify purpose* ▷

What crucially affects the answer to the manager's dilemma is the *ratio* between the profit on Kreemy and the profit on Yummy. This ratio was 30/10, or 3/1. However, suppose that when the Yummy company found it was getting no orders from this store and it discovered from the manager that this was because the profit on Yummy was too low, it increased the profit to 20p per Yummy carton, making the profit ratio now 30/20, or 3/2.

◁ *Create model* ▷

The expression for the overall profit in equation (27) would be replaced by:

$$P = 30x + 20y \qquad (28)$$

Some P-lines corresponding to equation (28) are shown in Figure 40.

◁ *Do the maths* ▷

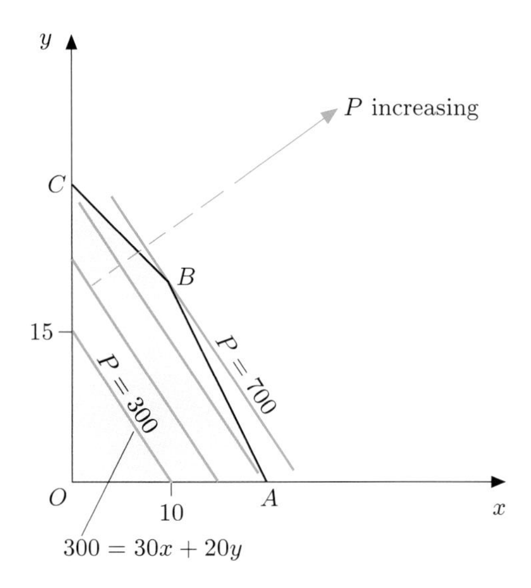

Figure 40

The line $P = 300$ is singled out in the figure—it cuts the axes at $x = 10$ and $y = 15$. You can see that as these parallel P-lines move away from the origin, the furthest you can go while staying at least partly within the feasible region is the point B. This has coordinates $x = 10$, $y = 20$.

So the manager's best strategy now is to stock 10 Kreemy and 20 Yummy cartons. In this case the sales constraint of 30 is fully met *and* the storage capacity is fully utilized. The total profit is now

◁ *Interpret results* ▷

$$30 \times 10 + 20 \times 20 = 700\text{p}$$

which is more than before.

One special feature of the solutions found to the problems discussed so far is that the solutions are at corners, or *vertices*, of the feasible region, namely A in Figure 39 and B in Figure 40. The *optimal solution*—that is, the one which produces the biggest profit—is usually at a vertex of the feasible region.

Activity 33 More ice cream

Suppose the Kreemy Company, on finding that its sales at the store dropped by half, increased the profit to the manager to 40p per carton, so the overall profit becomes:

$$P = 40x + 20y$$

Sketch some P-lines for this function and find an optimal solution.

In Activity 33 you probably noticed that the P-lines were parallel to the line AB. The line AB and the P-lines have the same *slope*. You can also see this from the fact that the ratio of the two profits is now $40/20 = 2/1$, the same as the ratio of the coefficients of x and y in the equation $x + \frac{1}{2}y = 20$ of the straight line AB. This time there is no unique solution to the problem—as the P-line moves in the direction indicated by the arrow in Figure 53, it is furthest away from the origin when it *coincides* with the line AB. In other words, *any* point on the line AB (with integer coordinates, because in this case x and y must be whole numbers) gives the maximum possible profit. Some examples are shown in Table 17.

Table 17

	x	y	$P = 40x + 20y$
Point A	20	0	800
Point B	10	20	800
A point in between A and B	12	16	800

▶ Do the points on AB where $x = 14, 16, 18$ give optimal solutions?

The points are $(14, 12)$, $(16, 8)$ and $(18, 4)$; each is integer-valued and so gives an optimal solution.

So, in certain circumstances, the optimal solution may not occur at a vertex, and indeed may not even be unique. However, in problems where there are more than two variables and constraints, the optimal solution almost invariably *does* occur at a vertex, except in very special cases; it is usually unique as well. However, such problems require a computer to solve them.

As was noted above, a solution of the ice-cream problem is only meaningful if the values of x and y are integers—the manager cannot stock fractions of a carton. How fortunate, then, that in all cases there were integer solutions. This cannot always be expected to happen, however; so, often, there needs to be an additional condition that x and y must be (non-negative) integers. Adding this further constraint to a linear programming problem turns it into an *integer programming* problem, which is harder to solve in general, as you may have to look at integer points near the optimal vertex but within the feasible region.

To summarize, linear programming problems are concerned with the allocation of limited resources in order to maximize a profit function (or minimize a cost function). The constraints, which represent the limitations on the resources, are described by linear inequalities. Together with the condition that no variables can be negative, in the two-variable case (x and y) this defines a feasible region bounded by straight lines in the upper right quadrant. The profit is represented by straight lines, and achieves a maximum value on the line furthest away from the origin that still lies partly within the feasible region. This usually occurs at a vertex of the feasible region.

Now work through Section 10.2 of Chapter 10 of the Calculator Book.

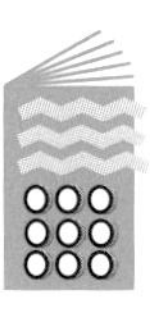

If you are not confident about tackling linear programming problems using your calculator then the audio band may help you. It tackles Example 14 below using the course calculator.

Example 14 High fashion

◁ *Specify purpose* ▷

A small-scale garment manufacturer makes two styles of ladies' coat. The first uses $1\,\text{m}^2$ of polyester material and $3\,\text{m}^2$ of wool material, whereas the second has a different, longer, style and requires $3\,\text{m}^2$ of polyester material and $2\,\text{m}^2$ of wool material. Each week the manufacturer receives $80\,\text{m}^2$ of polyester material and $120\,\text{m}^2$ of wool material. A wholesaler will accept any number of garments from the manufacturer, and pays £40 each for short coats and £60 each for longer coats. How many coats of each style should the manufacturer make in order to maximize the weekly income?

Now, if you wish, listen to band 1 of CDA5510 (Tracks 1–6).

AUDIO

◁ *Create model* ▷

The two variables in this problem are the number x of short coats and the number y of longer coats that the manufacturer produces each week.

The resources used can be expressed in tabular form, as in Table 18.

Table 18

	Short coats x	Longer coats y	Total available
Polyester (m^2)	1	3	80
Wool (m^2)	3	2	120
Income (£)	40	60	

The column headed 'short coats' gives the amounts (in m^2) of the materials needed per coat, and the income (in £) is listed at the bottom. The second column does the same for 'longer coats'. Reading across the rows gives the amounts of cloth used, so the total for polyester is $1 \times x + 3 \times y$. This cannot exceed the total available ($80\,\text{m}^2$), so the first constraint is:

$$x + 3y \leq 80$$

Similarly, reading across the second row of the table shows you that the amount of wool cloth used is $3x + 2y$. Since the total available is $120\,\text{m}^2$, the second constraint is:

$$3x + 2y \leq 120$$

As it is impossible to make negative numbers of coats, the variables must also satisfy the conditions $x \geq 0$, $y \geq 0$.

◁ *Do the maths* ▷

The feasible region can now be sketched. An easy way to do this without using your calculator is to find where each boundary line cuts the axes.

The first line

$$x + 3y = 80$$

cuts the y-axis when $x = 0$, at $y = 80/3 = 26\frac{2}{3}$; and the x-axis when $y = 0$, at $x = 80$; and is the line CD in Figure 41.

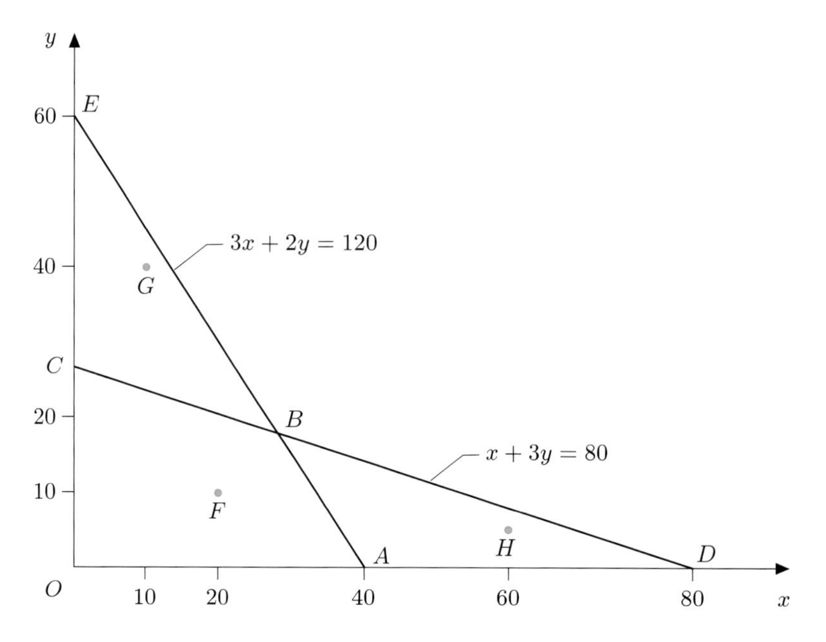

Figure 41

Similarly the second line

$$3x + 2y = 120$$

cuts the axes at the points E $(0, 60)$ and A $(40, 0)$.

To find the coordinates of the point B where these two lines meet, solve the simultaneous equations:

$$x + 3y = 80$$
$$3x + 2y = 120$$

You can plot the lines on your calculator to find the point of intersection or solve them algebraically. Point B has coordinates $x = 200/7 = 28\frac{4}{7}$ or approximately 28.6, $y = 120/7 = 17\frac{1}{7}$ or approximately 17.1.

▶ What is the feasible region?

The feasible region is region bounded by $OABC$. To check this, test appropriate points and find which ones satisfy both the inequality constraints. For example, the point F with $x = 20$, $y = 10$ has

$$x + 3y = 50 < 80$$
$$3x + 2y = 80 < 120$$

showing that F does lie inside the feasible region. However, the point G $(10, 40)$ gives

$$x + 3y = 130 > 80$$

and so lies outside the feasible region.

▶ Is H $(60, 5)$ inside the feasible region?

No. Although

$$x + 3y = 60 + 15 = 75 < 80$$

it is also true that:

$$3x + 2y = 180 + 10 = 190 > 120$$

◁ *Create model* ▷

Referring again to Table 18, you can see that the income from short coats is £$40x$ and from longer coats £$60y$, so the total weekly income in pounds is

$$P = 40x + 60y.$$

Some P-lines are drawn in Figure 42. The P-line furthest from the origin O passes through the vertex B, which therefore represents the optimal solution of $(28\frac{4}{7}, 17\frac{1}{7})$.

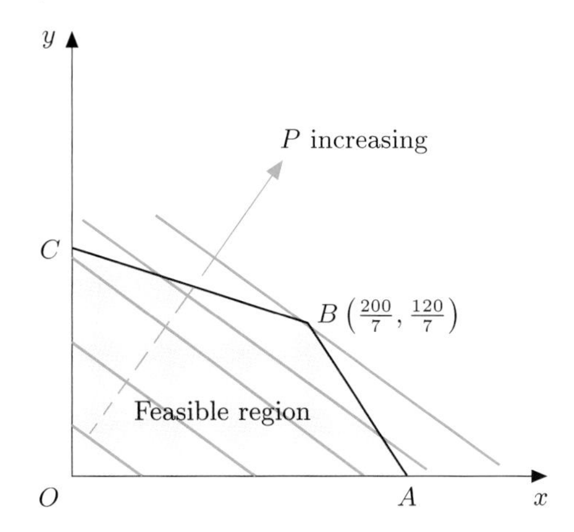

Figure 42 Feasible region for Example 14

◁ *Interpret results* ▷

There seems to be a difficulty: how can the manufacturer produce fractions of a coat? Perhaps any partially finished coats can be completed the following week. Over a seven-week period, the total numbers of short and longer coats produced will be seven times the weekly totals, that is 200 and 120 respectively. The total income over this seven-week period is

$$40 \times 200 + 60 \times 120 = £15\,200.$$

Alternatively, if weekly numbers are required, round the coat numbers down to the integer below—to $(28, 17)$, a point inside the feasible region. This gives a total weekly income of

$$40 \times 28 + 60 \times 17 = £2140.$$

Activity 34 Healthy living

Many people take vitamin and mineral tablets to supplement their diet. For those who eat a lot of 'fast food', such supplements are probably necessary: and, even if you try to maintain a well balanced diet, quite a lot of vitamin and mineral content can be lost in the preparation and cooking of foods. A regular supplement of vitamins and minerals may well do some good, and is very unlikely to do much harm! Choosing which tablets or capsules are the best to buy is a real problem, as a look at a list of ingredients on the packaging may confirm.

Suppose that you are unable to eat dairy products such as cheese or milk, and you decide to select a couple of tablets in order to keep up your intake of calcium, magnesium and zinc. The amounts (in milligrams) of these minerals in two multi-formula tablets are shown in Table 19.

Table 19

	Supervit	Maxivit	Minimum daily dose
Zinc (mg)	10	7	50
Magnesium (mg)	5	8	30
Calcium (mg)	30	10	90
Cost per tablet (pence)	12	9	

After reading through much literature on the beneficial effects of these minerals, you have decided you wish to receive a daily dose of at least 50 mg of zinc, 30 mg of magnesium and 90 mg of calcium. Supervit tablets cost 12p each and Maxivit cost 9p each. Use linear programming to determine how can you satisfy your requirements as cheaply as possible. (Be careful in your identification of the feasible region, and remember that you want to *minimize* the cost.)

At this stage you may well be feeling that you could have found the cheapest solution to Activity 34 simply by trying various integer combinations until you hit on the cheapest one which satisfied the constraints in the table. You would not have needed to draw any diagram, or do any algebra, eventually to have come up with the cheapest daily dose of *three* of each type of tablet. This gives a total daily cost of:

$$12 \times 3 + 9 \times 3 = 63\text{p}$$

It is certainly true that just trying out various guesses would work in this relatively simple problem, where there are only two unknown variables. If you added in another type of tablet, and perhaps one or two more minimum daily requirements for vitamins, you would very soon get into a hopeless muddle merely trying to find a solution that works (that is, which is inside the feasible region), never mind finding the cheapest mix.

Real-life problems are much more complicated, and are solved using computers. For example, a commercially used linear programming model (developed at Exeter University) is used to determine the best and cheapest diet for pigs at various stages of their growth. Pigs have minimum daily requirements of protein, carbohydrate, vitamins and minerals, and all together the problem involves twenty variables and ten constraints. Animal food companies and pig farmers use a computer program which solves this complicated problem. Imagine trying to do it all by guesswork—drawing diagrams using twenty variables is out of the question! Other areas where linear programming is widely used include the oil refining industry, where there are hundreds of variables and tens of thousands of constraints. Also, large communications networks apply linear programming to maximize their profitability: deciding how to allocate resources of long-distance cables, satellite stations and microwave transmitters in order to route millions of daily telephone and fax messages in the cheapest way is really big business!

Although this section has dealt with linear constraints, non-linear constraints sometimes arise. The principle of shading areas represented by such inequalities is the same. For instance, Figure 43 represents the region defined by the following constraints:

$$y \leq \sin x$$
$$y \geq 0$$
$$0 \leq x \leq 3$$

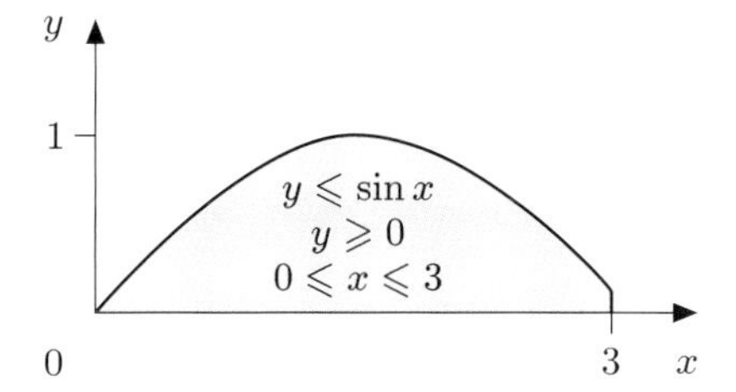

Figure 43

Activity 35

Draw on a graph the region represented by the following constraints:

$$y \leq \sin 2x$$
$$y \geq 0$$
$$0 \leq x \leq 1$$

You can either draw a sketch by hand or use your calculator.

Review your learning for this section by looking back through your Handbook entries for this section. Think also about how you 'read' the text.

As a result of your work on this section, you may now add inequalities as well as equalities to your library of mathematical models. Graphically an inequality may be represented by a region on one side of the line or curve representing the associated equality.

In commerce many constraints are expressible in the form of inequalities and the technique of linear programming may be useful in predicting the optimal strategy in such cases. In a two-variable situation, graphical methods may suffice. However, for more variables, computers are needed.

The procedure for solving a two-variable linear programming problem may be summarized as follows.

Solving a two-variable linear programming problem

(a) Represent each constraint by a straight line on a graph.

(b) Identify the area which satisfies all constraints. This is called the feasible region.

(c) Draw in some lines for the optimization (or profit) function and hence identify the optimal vertex.

(d) Find the coordinates of the vertex (either algebraically or graphically).

(e) Interpret the solution and look at adjacent integer points within the feasible region if appropriate.

Outcomes

After studying this section, you should be able to:

◇ identify regions on a graph corresponding to a given inequality (Activities 34, 35);

◇ use your calculator to display such areas (*Calculator Book*, Chapter 10, Section 10.2);

◇ solve linear programming problems involving two variables (Activities 33, 34).

Unit summary and outcomes

This unit describes the modelling process and the use of graphical and algebraic linear models, involving both equalities and inequalities.

A model is a representation of some aspect of reality created for a specific purpose. The main aspects of mathematical modelling are as follows.

Specify the purpose
Create a model
Do the mathematics
Interpret the results
Evaluate the model

The stages involved in the process of mathematical modelling are intended to help analyse and construct models. They comprise, however, a simplification of the modelling process, which is not always so logical in practice. The process is often called the mathematical modelling cycle, because initially a very simple model may be used which is later refined—by repeating some of all of the stages—until it meets the required specifications.

Activity 36 Reviewing ways of working

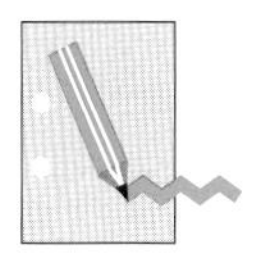

(a) Look back over your Handbook entries for this unit. Do you need to include any other terms or techniques, or add anything more to your entries?

(b) Read through any notes you have made on mathematical modelling. Do they make sense? Have they helped you to clarify the different types of models and the stages involved in developing them?

(c) Throughout this unit, you have been encouraged to develop a systematic approach to solving problems using the modelling stages. Think about how you have used these stages as you worked through the unit. Have you been labelling your own work with the stages as you worked through different problems? Is this a helpful technique for you?

(d) Think again about how you have been 'reading' this unit, and about the value of producing a running commentary (as if to a friend) on selected pieces of work.

(e) And, finally, look back at your study plan for completing this unit. How did it stand up to the actual workload? Do you need to make amendments to your study plan for the rest of this block? Record any changes on your study plan for the block.

The main ideas in this unit regarding the creation and solving of models, and the interpretation of results, can be summarized as follows.

Creating and solving a model

When creating a model, there are three basic steps.

(a) Determine the relevant variables.

(b) Make any necessary assumptions (for example, by assuming a constant rate of increase).

(c) Specify the limits of validity of the model.

Graphical linear model

When creating a graphical linear model, there are four additional steps which are sometimes useful.

(a) Determine which is the dependent variable (vertical axis) and which is the independent variable (horizontal axis).

(b) Assume a constant rate of increase (gradient) and find a numerical value for it.

(c) Find an initial value (intercept) for the dependent variable.

(d) Use the intercept and gradient to plot various points and join them by a straight line.

Sometimes an alternative to steps (b) and (d) is to assume a constant rate of increase and then to draw a straight line through known data points.

A graphical linear model is a simple one, which is used simply by reading values off the graph, but it may be better and easier to use an algebraic linear model.

Algebraic linear model

When creating an algebraic linear model, there are five additional steps which are sometimes useful.

(a) Determine which is the dependent variable and which the independent variable.

(b) Assume a constant rate of increase (gradient) and find a numerical value for it.

(c) Find an initial value (intercept) for the dependent variable or use a symbol to represent it.

(d) Write down the linear equation in the form

$$\text{dependent variable} = \text{rate of increase} \times \text{independent variable} + \text{initial value}$$

(e) Substitute known values for the variables into the equation in order to find the initial value (if necessary) and to check that the equation is correct.

An algebraic linear model is used by substituting in values of one variable in order to find the corresponding values of the other. Such models are quicker and more accurate to use than graphical linear models.

Fitting a linear model

The method of least squares can be used to find the 'best fit' straight line (linear model) through a set of data points. This is called the regression line.

Simultaneous linear models

Sometimes a problem can be solved by modelling it using more than one linear model. The solution is often at the intersection of the straight lines, and can be found most easily by algebraic methods involving substitution.

Inequalities

As well as equalities, linear inequalities are useful models, especially for commercial constraints. Inequalities are represented by regions on a graph rather than lines.

Linear programming

With constraints modelled by linear inequalities, and a linear model for profit (or other quantities to be optimized), graphical methods can be used for two-variable linear programming problems. The process is as follows.

(a) Represent each constraint by a straight line on a graph.
(b) Identify the area which satisfies all constraints. This is called the feasible region.
(c) Draw in some lines for the optimization (or profit) function and hence identify the optimal vertex.
(d) Find the coordinates of the vertex (either algebraically or graphically).

Interpreting the results

Results need to be interpreted in light of the purpose of the model, the assumptions made and the limits of the model's validity. The results may need to be rounded up or down according to the context and a safety margin may need to be included. In the case of linear programming problems, adjacent integer points within the feasible region may need to be considered.

Outcomes

Now you have finished this unit, you should be able to:

◇ explain in your own words the meaning of the following terms: 'model', 'modelling process', 'modelling cycle', 'variable', 'independent variable', 'dependent variable', 'parameter', 'linear model', 'linear graph', 'linear equation', 'rate of increase', 'gradient', 'initial value', 'intercept', 'function', 'interpolation', 'extrapolation', 'regression line', 'error bound', 'equilibrium price', 'simultaneous equations', 'inequality', 'linear programming', 'feasible region', 'constraint', 'optimal solution';

◇ distinguish the stages of the mathematical modelling cycle and use them when modelling;

◇ clarify the purpose of given modelling examples;

◇ find the gradient and intercept of a given straight-line graph;

◇ find the equation of a straight line;

◇ create single or simultaneous graphical or algebraic linear models and solve them;

◇ use your calculator to obtain a 'best fit' regression line for a given set of data points;

◇ interpolate or extrapolate, and describe the advantages and dangers associated with each;

◇ recognize and use scientific notation;

◇ use your calculator to sketch inequality constraints;

◇ solve two-variable linear programming problems;

◇ interpret the results from the various types of linear model encountered in this unit, and evaluate the suitability of such models for their purpose.

Comments on Activities

Activity 1

Being able to plan, prioritize and monitor your own progress is an important skill in life as well as for studying. The question you might ask yourself is 'Does it help my progress to plan and monitor what I'm doing?' For most people the response is 'Yes it does', however they do it.

Some people find this skill extremely difficult—it is also difficult to advise people since planning can happen in many different ways. If you feel you need more help with this skill try to discuss it with your tutor or other students. Try to persevere since being well organized and being able to plan and prioritize are things that will pay dividends.

If you are able to keep a record of your plans and how you have monitored your progress, you can use it as evidence for core skill achievement.

Activity 2

Here is one possible explanation.

A mathematical model is a mathematical representation of some aspect of reality created for a specific purpose—for instance, predicting how long a particular journey might take. It is sometimes useful to consider the process of creating and using a model in terms of the stages involved: specify the purpose; create a model; do the mathematics; interpret the results; and evaluate the model. If the evaluation suggests improvements, you can go through the stages again before using the results. So the process is sometimes called a modelling 'cycle'.

Activity 3

(a) quadratic model

(b) linear model

(c) sinusoidal model

Activity 4

(a) You may have described the graphs in different ways—here are some possible descriptions.

The graph of $y = x$ is a straight line.

$y = x^2$ is a curve—it comes down, turns round and goes up again, almost like a pudding basin.

$y = \sin x$ is a curve like an S on its side, or a picture of a wave, or a valley and a hill.

(b) The graphs of $y = \sin x$ and $y = x$ are very close near the origin. So $y = x$ is a useful approximation for $\sin x$ when x is close to zero.

(c) All the graphs look like straight lines.

Activity 5

(a) Put $S_1 = 50$ and $S_2 = 25$ into equation (1). Then:

$$\text{total time} = \left(\frac{20}{50} + \frac{10}{25}\right) \times 60 = 48 \text{ minutes}$$

(b) Here is one possible explanation.

The graphical model gave me a picture of what was happening, and of how the time taken would change with different assumptions about speed, because the gradient (or steepness) of the lines depends upon the speed of travel. However, the algebraic model, in the form of equation (1), enabled me to try out different assumptions more quickly and come up with a range of likely journey times from about 30 minutes (optimistic) to just under 50 minutes (pessimistic), and from this I realized that, unless we get going soon, we might miss the start of the tutorial by quite a lot.

Activity 6

Learning mathematics is a very complicated matter and there are many different ways of doing so. One of these is learning through reading, which OU students need to be able to do very effectively. But a lot of mathematics really has to be learning not only by reading text and but also by working through it. What does this mean? Mostly it means *you* taking the initiative with the text, that is, doing something with it. It could be working through the activities, checking the calculations, 'attacking' each paragraph, re-expressing what the text is saying, and so on. It is also important to see how any new material connects with what you already know. For example, were you able to see any connection between the statistical investigation stages and the modelling stages introduced in this section?

It may be useful to discuss with other students, friends or family how they 'learn' from a text and talk about how you have been 'reading' these units.

The essential points of the section are summarized in the main text following this activity.

Activity 7

There is no comment on this activity.

Activity 8

(a) (i) The relevant variables are the distance travelled, say d km, and the time taken since leaving port, say t hours (you might have used different letters).

(ii) The average speed of the ferry is:

$$\frac{40\,\text{km}}{2\,\text{h}} = 20\,\text{km per hour}$$

(iii) The speed is also $\dfrac{d\,\text{km}}{t\,\text{h}}$, so $\dfrac{d}{t} = 20$; this may be rearranged as follows:

$$d = 20t$$

(iv) The journey takes 2 hours; so the limitations are:

$$0 \leq t \leq 2$$

(b) $d = 20t \quad (0 \leq t \leq 2)$.

(i) $t = \frac{1}{4}$ gives $d = 20 \times \frac{1}{4} = 5$;

$t = \frac{70}{60}$ gives $d = 20 \times \frac{70}{60} = \frac{70}{3} = 23\frac{1}{3}$.

So after $\frac{1}{4}$ hour the ferry has travelled 5 km and after 1 hour and 10 minutes it has travelled just over 23 km.

(ii) When $d = 15$, $15 = 20t$ and $t = \frac{15}{20} = \frac{3}{4}$.

So the ferry is 15 km from England after $\frac{3}{4}$ hour.

When $d = 35$, $35 = 20t$ and $t = \frac{35}{20} = \frac{7}{4} = 1\frac{3}{4}$.

So the ferry is 35 km from England after $1\frac{3}{4}$ hours.

Activity 9

(a) Time (in hours) and depth of snow (in centimetres).

(b) Assume that the depth of snow increases at a constant rate of $19/10 = 1.9$ cm per hour. (Hence the model will be linear.)

(c) Points to plot are:

t (time after midnight, hours)	0	10
d (depth, cm)	0	19

These points can be joined by a straight line (as the model is linear).

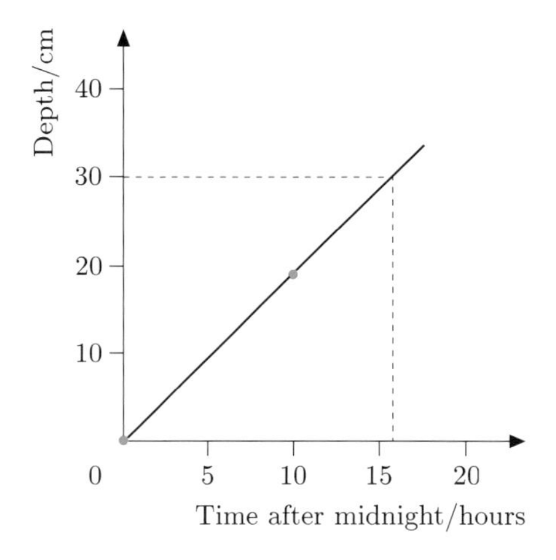

Figure 44

From the graph, the snow will be 30 cm deep at about 15.45 (3.45 pm).

(d) One answer is to tell the staff to start sufficiently before 15.45 (3.45 pm) in order to be able to clear the roads before the snow reached 30 cm, perhaps at 3.00 pm. There are other equally good interpretations leading to alternative strategies: for example, call out one person to inspect the situation at 2.00 pm and then make a final decision.

Activity 10

(a) The gradient is $\frac{10-5}{2-0} = 2.5$.
The intercept is 5.
The equation is $s = 2.5t + 5$.

(b) The gradient is $\frac{2-0}{20-0} = 0.1$.
The intercept is 0.
The equation is $d = 0.1t$.

(c) The gradient is $\frac{2-0}{3-1} = 1$.
The intercept is -1.
The equation is $x = t - 1$.

Activity 11

(a) The gradient is -1 and the intercept is 5, so the equation is:

$$C = -t + 5 \quad (\text{or } C = 5 - t)$$

(b) The gradient is -0.1 and the intercept is 1, so the equation is:

$$x = -0.1t + 1 \quad (\text{or } x = 1 - 0.1t)$$

(c) The gradient is -1 and the intercept is 0, so the equation is:

$$d = -t$$

(d) The gradient is -1 and the intercept is -1, so the equation is:

$$y = -x - 1$$

Note that, if you know two points on a linear graph, you can find the increase in both variables between the two points, and hence deduce the gradient of the graph.

Activity 12

(a) The gradient is $\frac{1}{2}$ and the intercept is 4, so the equation is:

$$y = \tfrac{1}{2}x + 4$$

(b) The gradient is $\frac{10-20}{3-1} = -5$ and so the equation is of the form

$$Q = -5P + c$$

where c is the intercept. Since the line passes through the point $(1, 20)$, it follows that $20 = -5 \times 1 + c$, so $c = 25$. Therefore the equation is:

$$Q = -5P + 25$$

Check that the point $(3, 10)$ satisfies the equation: $10 = -5 \times 3 + 25$.

(c) The gradient is $\frac{0-2}{2-1} = -2$ and so the equation is of the form

$$d = -2t + c$$

where c is the intercept. The line passes through the point $(2, 0)$, so $0 = -2 \times 2 + c$, giving $c = 4$. Therefore the equation is

$$d = -2t + 4.$$

Check that the point $(1, 2)$ satisfies the equation: $2 = -2 \times 1 + 4$.

Activity 13

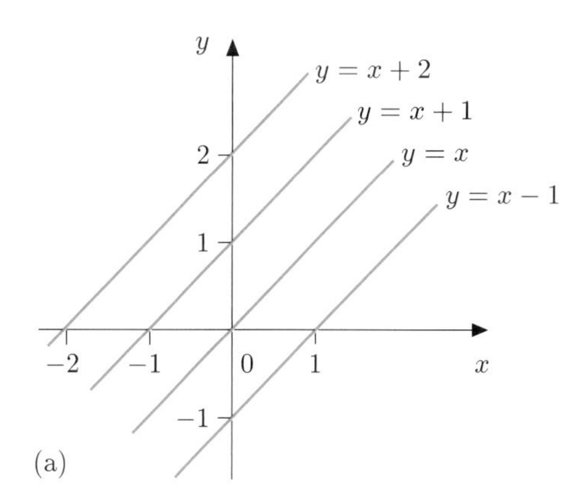

(a)

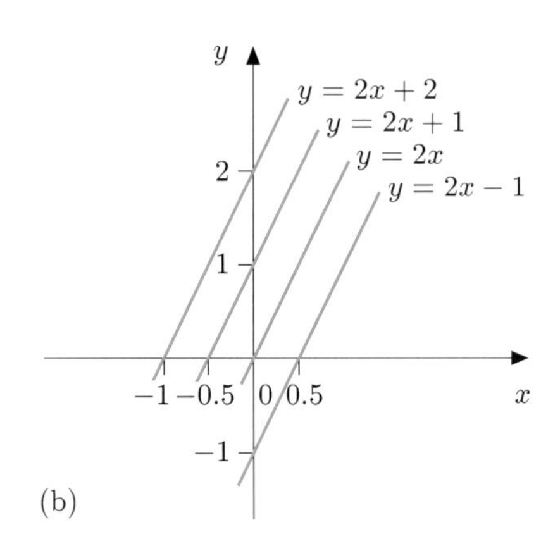

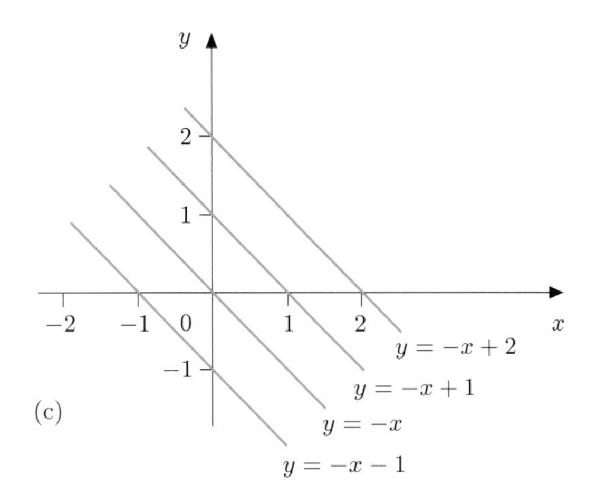

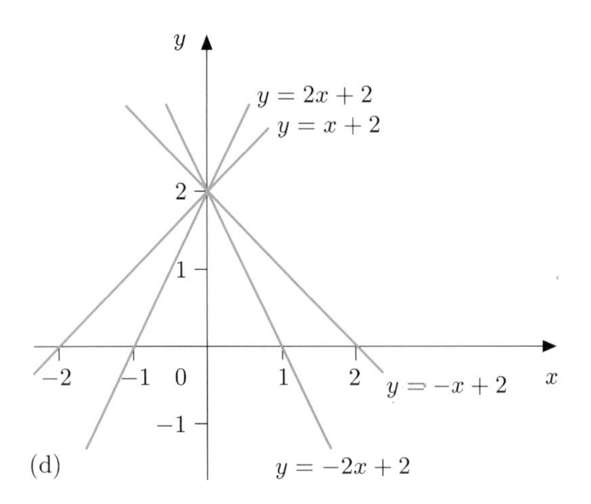

Figure 45

(e) Graphs of equations of the form $y = mx + c$ with the same value of m consist of parallel lines (because they have the same gradient). When the equations all have the same value of c, their graphs consist of straight lines that all pass through the same point—the intercept on the y-axis, $(0, c)$.

Activity 14

It would be more sensible to suggest that they start snow-ploughing at 15.45; or to suggest 'between 15.30 and 16.00', as the model might not be accurate to within more than half an hour or so.

Activity 15

(a) Assume that there is a linear relationship between quantity sold and price, plot the points on a graph and draw a straight line through them, as in Figure 46.

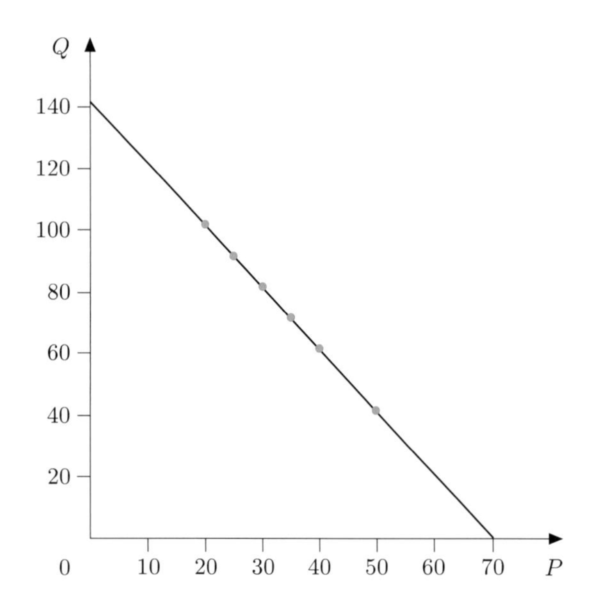

Figure 46

(b) (i) about 50 kg per day

(ii) about 112 kg per day

(iii) about 86 kg per day

(iv) about 6 kg per day

(c) In cases (i) and (iii), where the results are obtained by interpolation and can be expected to be reasonably accurate, it is probably best for the greengrocer to buy as near as possible to the given number of kilograms (she may not be able to buy in less than 10 kg lots).

However, in case (iv), the result is likely to be rather inaccurate, as the line is extrapolated a long way beyond the data points. Therefore she should probably buy the smallest quantity allowed, if any at all.

(A model which tails off, as in Figure 47, is likely to be more realistic than a linear model for prices at the extremes of the range.)

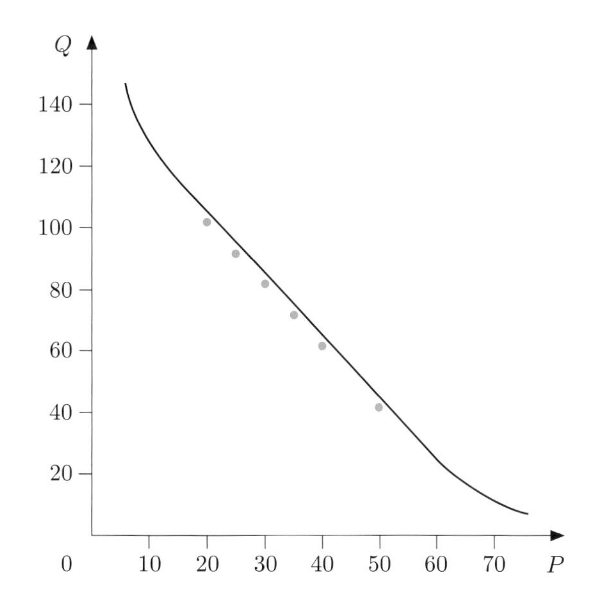

Figure 47

Activity 16

(a) The model is only valid for a motorway journey between Nottingham and Milton Keynes and for $1\frac{1}{2}$ hours. It would also not be valid if there were any hold-up.

(b) The model is only valid while it continues to snow at roughly the same rate.

(c) The model is only valid until the river bursts its banks ($d = 26$) and while the water rises at roughly the same rate.

Activity 17

(a) Take the number of instruments made per week to be N and the corresponding total costs to be £C. Assume that C increases at a constant rate with N, which leads to the linear graph in Figure 48.

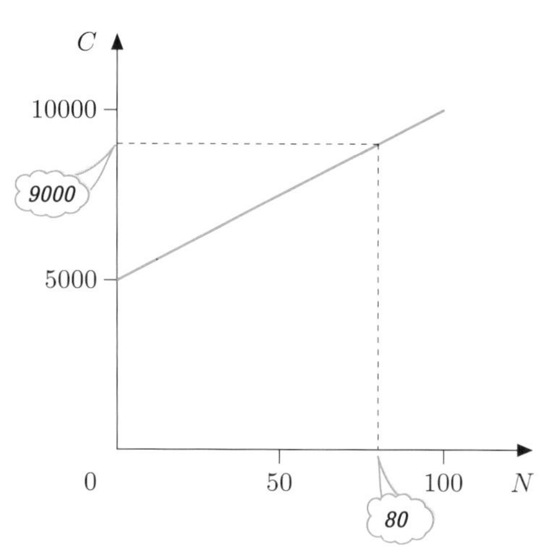

Figure 48

The intercept is 5000 and the gradient is 50. So the equation of the line is:

$$C = 50N + 5000$$

(b) The model is only valid for N between 0 and 100 as the firm has the capacity to produce only up to 100 instruments per week:

$$0 \leq N \leq 100$$

(c) The intercept represents the costs which are incurred even if no instruments are made—the fixed costs. The gradient gives the cost per instrument over and above these fixed costs (this is sometimes referred to as the 'marginal cost' or 'variable cost').

(d) If $N = 80$, then:

$$C = 50 \times 80 + 5000 = 9000$$

So the cost would be about £9000.

Activity 18

(a) Let the quantity which it is economical to supply be Q kg, at a price of P pence per kg.

Assume a constant rate of increase of Q with P. This rate is:

$$\frac{1600 - 800}{50 - 30} = 40$$

The algebraic linear equation is therefore of the form

$$Q = 40P + Q_0$$

where Q_0 is the initial value.

This is valid for P in the region of about 20 pence to 60 pence per kg. (You may have been more cautious and said '25 pence to 55 pence per kg'. There is no one 'right' answer.)

Q is 800 when P is 30, so $800 = 40 \times 30 + Q_0$, giving $Q_0 = -400$. Hence the algebraic linear model is:

$$Q = 40P - 400$$

Check that the point $P = 50$, $Q = 1600$ satisfies the equation:

$$1600 = 40 \times 50 - 400$$

(b) When $P = 35$, $Q = 40 \times 35 - 400 = 1000$. So it would be economical to supply 1000 kg.

When $P = 40$, $Q = 40 \times 40 - 400 = 1200$. So it would be economical to supply 1200 kg.

When $P = 55$, $Q = 40 \times 55 - 400 = 1800$. So it would be economical to supply 1800 kg.

Activity 19

You should have identified the following terms, and maybe others:

variables and parameters;
linear or straight-line model, graph and equation;
gradient or rate of increase;
intercept or initial value;
dependent and independent variable;
function, its range or domain;
interpolation and extrapolation.

If you are unsure of the meaning of any of them, go back over the section and reread the appropriate part(s) of the text.

Specialist terms should only be used when the audience understands them. If necessary you should explain the meaning of any specialist terms you need to use.

Activity 20

Every student's description will be different. Your description should, however, include: the notion of 'scatter'; the idea of a 'useful' (rather than 'correct') straight-line model; and the idea of the 'best' line identified by minimizing the sum of the squared point-to-line distances.

Activity 21

(a) The calculator gives:

$$a = -2$$
$$b = 142$$
$$r = -1$$

So the regression line is $y = -2x + 142$.

The gradient $a = -2$ indicates the rate of increase of quantity sold with price (actually a decrease of 2 kg in sales for every 1p increase in price).

The intercept $b = 142$ indicates the quantity 'sold' when the price is zero (the 'giveaway' price).

The correlation coefficient $r = -1$ indicates perfect correlation (all the data points lie on the regression line) and a negative slope.

(b) The calculator gives:

$$a = 58$$
$$b = 0$$
$$r = 1$$

So the regression line is $y = 58x$.

The gradient $a = 58$ indicates the rate of increase of distance with time, that is the average speed (in mph).

The intercept $b = 0$ indicates the distance travelled when the time is zero (at the start of the journey).

The correlation coefficient $r = 1$ indicates perfect correlation (all the data points lie on the regression line) and a positive slope.

The equation $y = 58x$ of the regression line is the same as the equation $d = 58t$ derived in Subsection 2.2 (but using different letters for the variables).

(c) The calculator gives:

$$a = -20$$
$$b = 2200$$
$$r = -1$$

So the regression line is $y = -20x + 2200$.

The gradient $a = -20$ indicates the rate of increase in demand with price (actually a decrease of 20 kg in demand for every 1p increase in price).

The intercept $b = 2200$ indicates the demand when the price is zero (the 'giveaway' price).

The correlation coefficient $r = -1$ indicates perfect correlation (all the data points lie on the regression line) and a negative slope.

The equation $y = -20x + 2200$ of the regression line is the same as the equation $Q = 2200 - 20P$ derived in Example 6 (after rearrangement and using different letters for the variables).

Activity 22

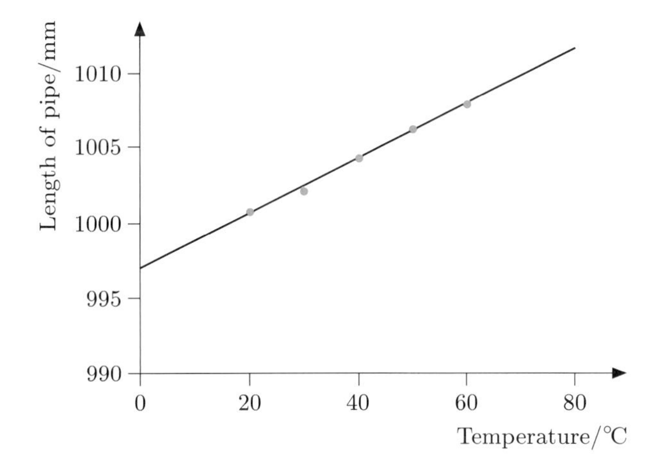

Figure 49

The equation of the line is:

$$L = 0.168T + 997.3$$

The intercept is 997.3 (to one decimal place): it is the predicted length of the tube in mm at 0 °C.

The slope is 0.168 (to three significant figures): it is the amount (in mm) by which the tube expands for each increase of 1 °C in the temperature.

Activity 23

It is much easier to use the metric measurements!

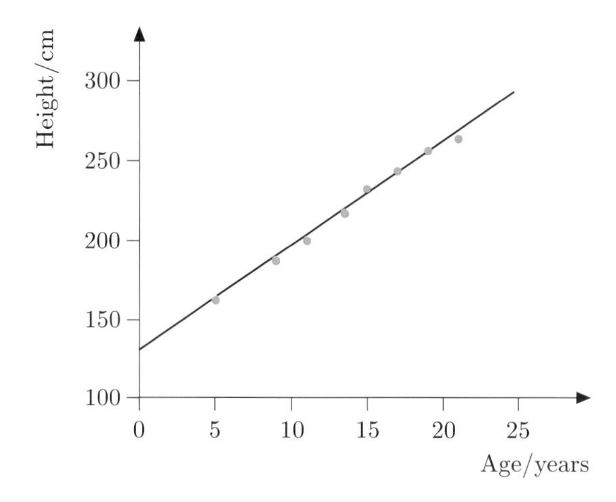

Figure 50

The equation of the line is:

$$y = 6.638x + 130$$

The average annual rate of growth is given by the gradient of the line, which is 6.638 (to three decimal places). So the annual rate of growth is about 6.6 cm per year.

Activity 24

Your summary will be personal to you. However, the paragraphs which follow in the main text summarize the argument. So check you have included all the main points in your summary.

Activity 25

(a) $6.696\,000 \times 10^6$

(b) 6.696×10^6

Activity 26

These measurements are personal to you, but you are unlikely to be able to measure your height more accurately than to the nearest 0.001 m (1 mm), your weight more accurately than to the nearest 0.1 kg or journey time more accurately than to the nearest minute.

Activity 27

(a)

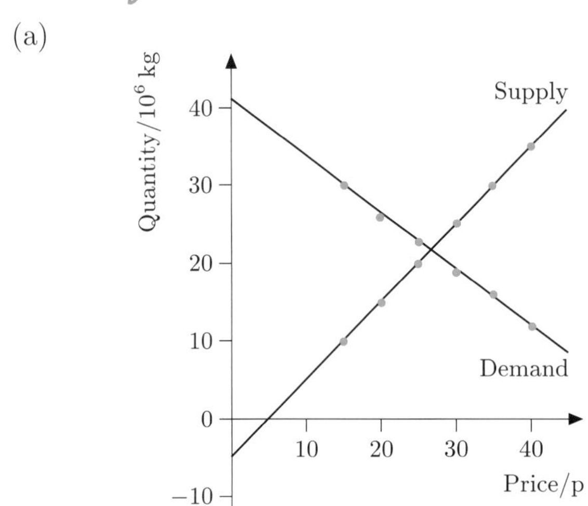

Figure 51

(b) The estimated price is about 27p, at which price about $21\frac{1}{2}$ million kg of fruit are sold.

Activity 28

Substitute from (7) into (8):

$$2(7+Y) = 2+Y$$
$$14+2Y = 2+Y$$
$$2Y = -12+Y$$
$$Y = -12$$

Substitute for Y in (7), giving:

$$X = 7-12 = -5$$

Check, using $Y = -12$ in (8):

$$2X = 2-12 = -10$$

which is correct.

Activity 29

You may have used different rearrangements and substitutions from the ones given here, but you should have obtained the same solutions.

(a)
$$2X = Y - 1 \qquad (13)$$
$$3X = Y + 1 \qquad (14)$$

From (13):

$$Y = 2X + 1$$

Substituting in (14) gives:

$$3X = 2X + 1 + 1$$
$$X = 2$$

Substituting for X in (13) gives:

$$2 \times 2 = Y - 1$$
$$Y = 5$$

Checking in (14) gives $3 \times 2 = 5 + 1$, which is correct.

So the solution is $X = 2$, $Y = 5$.

(b)
$$X = 2Y - 3 \qquad (15)$$
$$3X = -2Y + 15 \qquad (16)$$

From (15):

$$X = 2Y - 3$$

Substituting in (16) gives:

$$3(2Y-3) = -2Y + 15$$
$$6Y - 9 = -2Y + 15$$
$$8Y = 24$$
$$Y = 3$$

Substituting for Y in (15) gives:

$$X = 2 \times 3 - 3 = 3$$

Checking in (16) gives $3 \times 3 = -2 \times 3 + 15$, which is correct.

So the solution is $X = 3$, $Y = 3$.

(c) $$2A = 5B + 2 \qquad (17)$$
$$2A = 3B - 2 \qquad (18)$$

Substituting for $2A$ from (17) into (18):

$$5B + 2 = 3B - 2$$
$$2B = -4$$
$$B = -2$$

Substitution for B in (17) gives:

$$2A = 5(-2) + 2$$
$$2A = -8$$
$$A = -4$$

Checking in (18) gives $2(-4) = 3(-2) - 2$, which is correct.

So the solution is $B = -2$, $A = -4$.

(d) $$3P = 3 + 3Q \qquad (19)$$
$$7P = 1 + 4Q \qquad (20)$$

Dividing equation (19) by 3 gives:

$$P = 1 + Q$$

Substituting in (20) gives:

$$7(1 + Q) = 1 + 4Q$$
$$3Q = -6$$
$$Q = -2$$

Substituting for Q in (19) gives:

$$3P = 3 + 3(-2)$$
$$3P = -3$$
$$P = -1$$

Checking in (20) gives $7(-1) = 1 + 4(-2)$, which is correct.

So the solution is $Q = -2$, $P = -1$.

Activity 30

(a) Relevant variables are the total cost £C at time t months after the start of the project.

The model is valid for the length of the project, that is for $0 \leq t \leq 12$.

Option 1

Assume the total cost is made up only of the purchase price of £8000, the cost of two programmers at £1600 per month, the running costs of £30 per month, and maintenance costs of £120 per month. So

$$C = 8000 + 1600t + 30t + 120t$$

which simplifies to:

$$C = 8000 + 1750t \qquad (29)$$

Option 2

Assume the total cost is made up only of the purchase price of £1500, the cost of three programmers at £2400 per month, the running costs of £40 per month, and maintenance costs of £100 per month. So

$$C = 1500 + 2400t + 40t + 100t$$

which simplifies to:

$$C = 1500 + 2540t \qquad (30)$$

Option 3

Assume the total cost is made up only of the machine rental of £850 per month, the cost of two programmers at £1600 per month, running costs of £30 per month, and maintenance costs of £120 per month. So

$$C = 850t + 1600t + 30t + 120t$$

which simplifies to:

$$C = 2600t \qquad (31)$$

(b) *Options 1 and 3*

$$C = 8000 + 1750t \qquad (29)$$
$$C = 2600t \qquad (31)$$

Initially option 3 is more economical, as there is no initial cost.

There is a break-even point if there are values for C and t which simultaneously satisfy equations (29) and (31).

Using substitution gives:

$$8000 + 1750t = 2600t$$
$$8000 = 850t$$

So $t = \frac{160}{17} = 9.4$ months (to two significant figures).

Check in (29):
$C = 8000 + 1750 \times \frac{160}{17} = 24\,471$
(to five significant figures)

Check in (31):
$C = 2600 \times \frac{160}{17} = 24\,471$
(to five significant figures)

So for about the first $9\frac{1}{2}$ months option 3 is likely to be more economical, and after this option 1 is likely to be more economical.

(c) *Options 1 and 2*

$$C = 8000 + 1750t \quad (29)$$

$$C = 1500 + 2540t \quad (30)$$

Initially option 2 is more economical.

The break-even point comes when (29) and (30) are simultaneously satisfied.

Using substitution gives:

$$8000 + 1750t = 1500 + 2540t$$
$$6500 = 790t$$

So $t = \frac{650}{79} = 8.2$ (to two significant figures).

Check in (29):
$C = 8000 + 1750 \times \frac{650}{79} = 22\,399$
(to five significant figures)
Check in (30):
$C = 1500 + 2540 \times \frac{650}{79} = 22\,399$
(to five significant figures)

So for about 8 months option 2 is more economical, and after this option 1 becomes more economical.

Options 2 and 3

$$C = 1500 + 2540t \quad (30)$$

$$C = 2600t \quad (31)$$

Initially option 3 is more economical.

There is a break-even point if (30) and (31) are simultaneously satisfied.

Using substitution gives:

$$1500 + 2540t = 2600t$$
$$1500 = 60t$$

So $t = 25$.

Check in (30):
$C = 1500 + 2540 \times 25 = 65\,000$
Check in (31):
$C = 2600 \times 25 = 65\,000$

Since $t = 25$ is outside the range for this model (the project is expected to take 10 to 12 months), there will not be a break-even point during the length of the project. Hence option 3 is likely to be more economical than option 2 throughout the project.

So, overall, option 3 is most economical for the first $9\frac{1}{2}$ months or so and then option 1 becomes more economical.

(d) The assumptions involved ignoring any other factors (for example, interest on the money to buy the program development aid and any resale value it might have after the project). Such things as staff costs are also simplified, as there could be training costs and/or redundancy payments.

Overall this is very much a first model and more detailed models will be necessary before a final decision can be reached. However, it does look as though option 2 is not going to be the most economical and the firm might decide to drop this option now, unless there are other factors (such as the need to keep current staff employed or the future use of the aid) which might outweigh the economic factors. Also, on the basis of the information obtained so far, option 1 looks the most attractive choice over 10 to 12 months; but a more refined model is really required before a final choice can be made.

Activity 31

All the models are based on an assumption that the mean running velocity for each event is increasing linearly. The linear models for women have steeper gradients than those for men, and so if this trend continues women would catch men up. The simultaneous linear models for men and women running the same distances intersect at a point which predicts when women will run as fast as men and how fast this will be. This ranges from around 1995 for the marathon to 2055 for the 200 metres. However, the extrapolation of models so far is unreliable.

Activity 32

Let x be the number of Kreemy cartons and y the number of Yummy cartons that the manager buys each week. Since the total bought is 30, this means that:

$$x + y = 30 \quad (32)$$

The Kreemy cartons occupy 1 litre of space each and the Yummy cartons $\frac{1}{2}$ litre each. Since the total volume they take up in the cabinet is 20 litres, and assuming that *all* the storage space is used up, this means that:

$$x + \tfrac{1}{2}y = 20 \quad (33)$$

The manager's problem is to find the values of x and y that simultaneously satisfy the two equations.

From (32), $x = 30 - y$. Substitute into (33):

$$30 - y + \tfrac{1}{2}y = 20$$
$$10 = \tfrac{1}{2}y$$
$$y = 20$$

Hence from (32), $x = 30 - 20 = 10$.

Check in (33): $10 + \frac{1}{2} \times 20 = 20$, which is correct.

So the manager buys 10 cartons of Kreemy and 20 cartons of Yummy.

Another way of solving equations is to rearrange them into the form $y = f(x)$, plot them using your calculator, and find the point of intersection of the two straight lines, as in Figure 52.

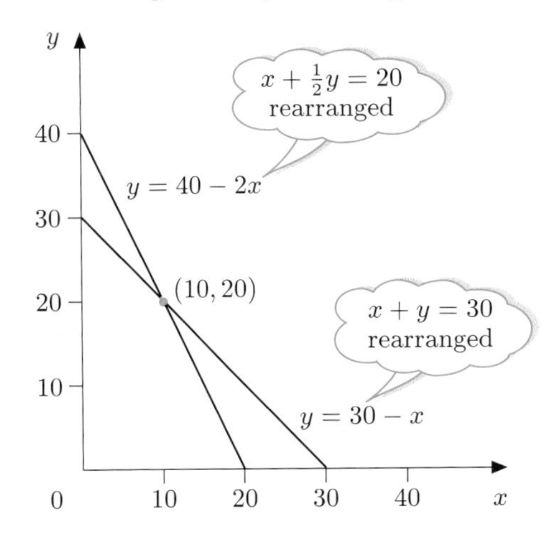

Figure 52

Activity 33

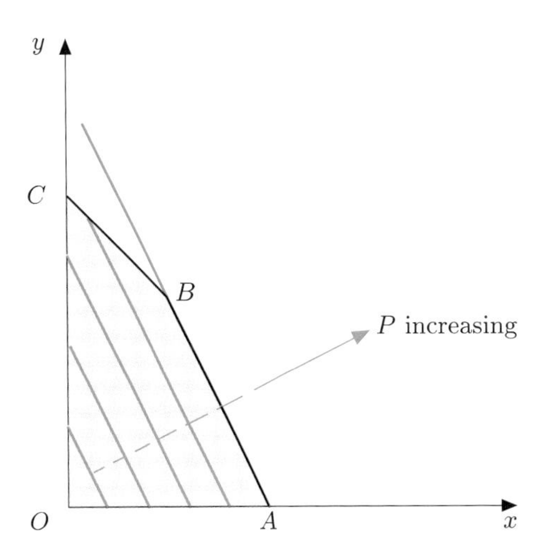

Figure 53

Any point on the line AB is an optimal solution.

Activity 34

Let x be the number of Supervit tablets and y the number of Maxivit tablets that you take each day.

Reading across the first row of the table shows that you will receive a daily dose of $10x + 7y$ milligrams of zinc. This total must be at least 50, so the first constraint is:

$$10x + 7y \geq 50 \quad (34)$$

(Notice that this inequality is 'the other way round' to those in Example 14. This is because the requirement on the zinc dose must be equalled *or exceeded*, whereas in the coat manufacturing problem the availability of cloth could *not* be exceeded.)

The straight line

$$10x + 7y = 50$$

cuts the axes at $x = 5$ and $y = 50/7 = 7\frac{1}{7}$, as shown in Figure 54.

Of course, you cannot take negative numbers of tablets, so as usual only the quadrant where $x \geq 0$, $y \geq 0$ needs to be considered.

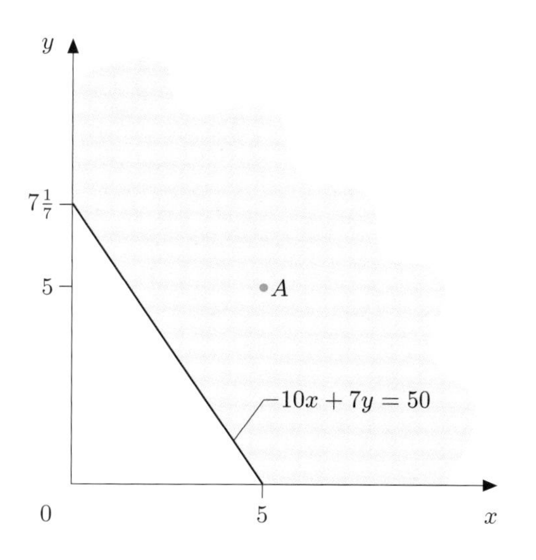

Figure 54

The point A in Figure 54, with $x = 5$, $y = 5$, is such that $10x + 7y = 85 > 50$, so the region represented by inequality (34), shaded in Figure 54, is above the line. (This is the opposite of what happened in Example 14, where the inside of the corresponding triangular region was the one represented by the inequality.)

Reading along the second and third rows of the table produces in the same way the following two constraints:

$$5x + 8y \geq 30 \quad \text{(magnesium requirement)}$$
$$30x + 10y \geq 90 \quad \text{(calcium requirement)}$$

The feasible region is shown in Figure 55.

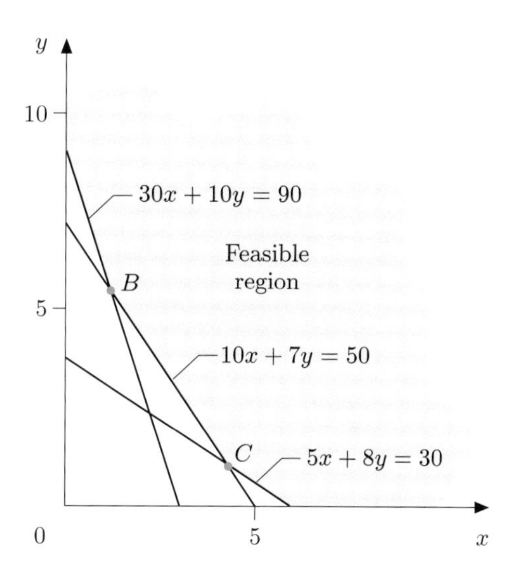

Figure 55 Feasible region

You can find the coordinates of the points B and C by solving the appropriate pairs of equations. For example, B is the point where the lines

$$10x + 7y = 50$$
$$30x + 10y = 90$$

intersect, and has coordinates $x = 1\frac{2}{11}$ or approximately 1.2, $y = 5\frac{5}{11}$ or approximately 5.5. Similarly, C has coordinates $x = 4\frac{2}{9}$ or approximately 4.2, $y = 1\frac{1}{9}$ or approximately 1.1.

The cost which you wish to *minimize* is:

$$P = 12x + 9y$$

The straight line $P = 84$ is shown in Figure 56. It cuts the axes at $(7, 0)$ and $(0, 9\frac{1}{3})$.

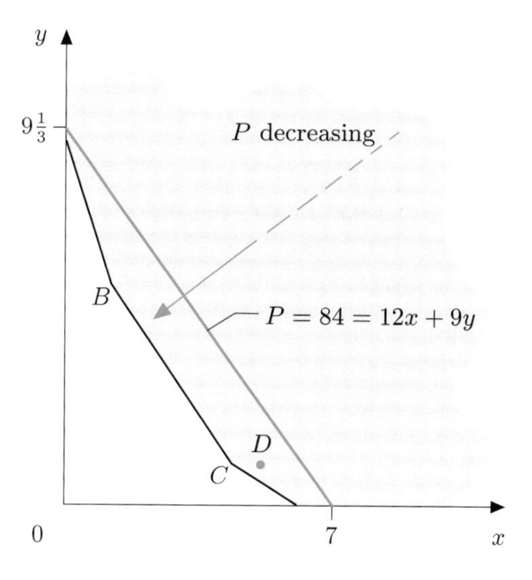

Figure 56

This time, since P is to be as *small* as possible, this is achieved when the P-line is as *close* to the origin as possible, while still at least partly in the feasible region. From Figure 56, you can see that this occurs when the P-line passes through the vertex C. So the optimal solution occurs at this vertex of the feasible region: at $(4\frac{2}{9}, 1\frac{1}{9})$.

The solution to the problem has not come out in integer form, and you cannot take $4\frac{2}{9}$ tablets of Supervit and $1\frac{1}{9}$ of Maxivit unless you have a very sharp knife and a keen eye for cutting up a ninth of a tablet! What you *can* do to implement this dose is to multiply each daily dose by nine, so as to make it an integer, and then take 38 ($= 9 \times 4\frac{2}{9}$) tablets of Supervit and 10 ($= 9 \times 1\frac{1}{9}$) tablets of Maxivit over a *nine-day* period. The cost over this period is

$$12 \times 38 + 9 \times 10 = 546\text{p}.$$

However, you might not want the fiddle of having to count up your tablets over a nine-day period, so you could decide instead to round off the doses to the nearest whole number giving 4 Supervit and 1 Maxivit tablets per day. However, this is unsatisfactory, since it does not satisfy the zinc requirement—putting $x = 4$ and $y = 1$ gives $10x + 7y = 40 + 7 < 50$; that is, the point $(4, 1)$ is outside the feasible region. (This illustrates one of the difficulties in linear programming when you require the solution to consist of integers.) Try instead $x = 5$, $y = 1$. This now satisfies the zinc requirement: $10x + 7y = 50 + 7 > 50$; and you can check that the other two constraints are satisfied also. So this 'solution' is to take five Supervit and one Maxivit tablet per day. However, there is a price to pay—literally—for using this more convenient daily dose. The total cost over a nine-day period is now

$$(12 \times 5 + 9 \times 1) \times 9 = 621\text{p}.$$

This costs more, as you are now receiving bigger doses of the three minerals: you can see this by looking at Figure 56, where D is the point $x = 5$, $y = 1$, which is not a vertex of the feasible region, but is inside it. You could try out other integer solutions within the feasible region near the vertex C in Figure 56, such as $x = 4$, $y = 2$, to see whether any of them is actually better (cheaper). You should find that $(3, 3)$ is optimal. So take three of each tablet per day, although this costs £5.67.

Activity 35

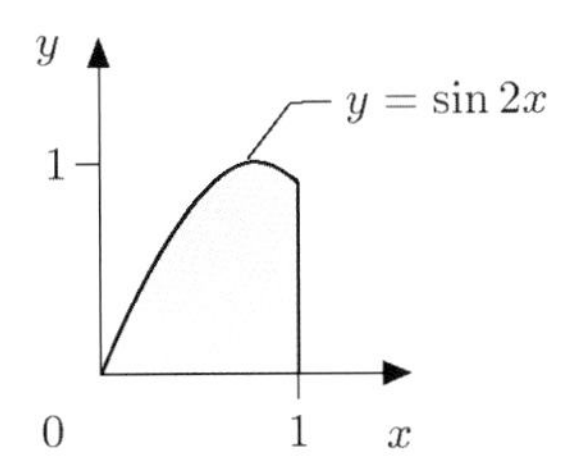

Figure 57

Activity 36

There is no comment on this activity.

Acknowledgements

Grateful acknowledgement is made to the following sources for permission to reproduce material in this unit.

Cover

John Regis, sprinter: Press Association; crowd scene: Camera Press; car prices: Edinburgh Mathematical Teaching Group; other photographs: Mike Levers, Photographic Department, The Open University.

Index